협소주택 1

작은 면적, 넓은 공간

협소주택 1

15m² - 25m²

ROOFTOP HOUSE ǀ 옥상공간을 활용하여 도시환경의 기회와 가능성을 담은 집	030 page
LONG WINDOW HOUSE ǀ 긴 창이 있는 집	044 page
HOUSE AT HOMMACHI ǀ 천창을 통해 환기와 채광을 극대화한 호마치 주택	056 page
PH LAVALLEJA ǀ 기하학적인 구조로 외관의 연속성을 살린 PH 라바예하	066 page
SLICE ǀ 정원의 한 조각 같은 예술적인 집	078 page
MINI HOUSE KODA ǀ 사전제작하여 배송하는 이동식 주거, 코다	084 page

25m² - 35m²

HOUSE IN KAWASAKI ǀ 대형 접이식문을 이용하여 실내외를 연결한 가와사키 주택	100 page
YEONNAM-DONG MIX-USE HOUSING ǀ 다양한 공간감을 가진 연남동 주택	112 page
HOUSE IN THE CITY ǀ 다층구조로 최대의 면적을 확보한 도시주택	124 page
TRANS ǀ 도시생활을 즐기는 젊은 커플을 위한 주택, 트랜스	140 page
H33617 ǀ 불편한 집과 행복한 집 사이, 적은집 H33617	152 page
NEST ǀ 도심 속 새둥지 같은 아늑한 소형주택, 네스트	160 page
[CREVICE] 1740 ǀ 거대한 도시 속 작은 땅, 새로운 기회, 틈 1740	170 page
SMALL HOUSE WITH FLOATING TREEHOUSE ǀ 떠있는 트리하우스를 담은 작은집	184 page
SIXTEEN ROOMS ǀ 거주자의 개성을 담은 16개의 작은방	194 page
Mrs. FAN'S PLUGIN HOUSE ǀ 천창을 통해 빛으로 가득 채운 판씨네 플러그인 하우스	206 page
SEONGSAN-DONG MIX-USE HOUSING ǀ 골목과의 유동적인 관계를 만든 집, 성산동 주택	218 page
ACUTE HOUSE ǀ 삼각형 모양의 대지를 활용한 어큐트 하우스	232 page
NARVARTE TERRACE ǀ 옥상공간을 활용한 소형주거, 나바르테 테라스	246 page

35m² - 41m²

CIRCULATE HOUSE ǀ 공간의 연속과 분할이 조화를 이루는 순환가옥	258 page
S1927 ǀ 스킵플로어로 공간을 연결한 적은집 S1927	272 page
HAT ǀ 자연광이 풍부하게 들어오는 집 HAT	280 page
H4912 ǀ 다양한 사이즈의 창들로 풍부한 조망을 제공하는 적은집 H4912	288 page
SUMMERHOUSE T ǀ 자연을 향한 개방형 구조를 통해 다양한 공간경험을 제공하는 썸머하우스T	298 page
Y HOUSE ǀ 연창을 통해 주변 자연을 받아들인 Y 하우스	308 page
VICOLO ǀ 공생의 건축을 담은 협소건물, 비꼴로	318 page
SUBAKO ǀ 테라스가 달린 집 수바코	332 page

CONTENTS_BY SIZE

Narrow-Urban

LONG WINDOW HOUSE ǀ 긴 창이 있는 집	044 page
HOUSE AT HOMMACHI ǀ 천창을 통해 환기와 채광을 극대화한 호마치 주택	056 page
HOUSE IN KAWASAKI ǀ 대형 접이식문을 이용하여 실내외를 연결한 가와사키 주택	100 page
TRANS ǀ 도시생활을 즐기는 젊은 커플을 위한 주택, 트랜스	140 page
NEST ǀ 도심 속 새둥지 같은 아늑한 소형주택, 네스트	160 page
SMALL HOUSE WITH FLOATING TREEHOUSE ǀ 떠있는 트리하우스를 담은 작은집 협소주택	184 page
CIRCULATE HOUSE ǀ 공간의 연속과 분할이 조화를 이루는 순환가옥	258 page

Angled-Corner

ACUTE HOUSE ǀ 삼각형 모양의 대지를 활용한 어큐트 하우스	232 page

Typical-Small

ROOFTOP HOUSE ǀ 옥상공간을 활용하여 도시환경의 기회와 가능성을 담은 집	030 page
HOUSE IN THE CITY ǀ 다층구조로 최대의 면적을 확보한 도시주택	124 page
H33617 ǀ 불편한 집과 행복한 집 사이, 적은집 H33617	152 page
[CREVICE] 1740 ǀ 거대한 도시 속 작은 땅, 새로운 기회, 틈 1740	170 page
SEONGSAN-DONG MIX-USE HOUSING ǀ 골목과의 유동적인 관계를 만든 집, 성산동 주택	218 page
HAT ǀ 자연광이 풍부하게 들어오는 집 HAT	280 page
H4912 ǀ 다양한 사이즈의 창들로 풍부한 조망을 제공하는 적은집 H4912	288 page
SUBAKO ǀ 테라스가 달린 집 수바코	332 page

Compact-Nature

SLICE ǀ 정원이 한 조각 같은 예술적인 집	078 page
SUMMERHOUSE T ǀ 자연을 향한 개방형 구조를 통해 다양한 공간경험을 제공하는 썸머하우스T	298 page
Y HOUSE ǀ 연창을 통해 주변 자연을 받아들인 Y 하우스	308 page

Multi-use

PH LAVALLEJA ǀ 기하학적인 구조로 외관의 연속성을 살린 PH 라바예하	066 page
MINI HOUSE KODA ǀ 사전제작하여 배송하는 이동식 주거, 코다	084 page
YEONNAM-DONG MIX-USE HOUSING ǀ 다양한 공간감을 가진 연남동 주택	112 page
SIXTEEN ROOMS ǀ 거주자의 개성을 담은 16개의 작은방	194 page
S1927 ǀ 스킵플로어로 공간을 연결한 적은집 S1927	272 page
VICOLO ǀ 공생의 건축을 담은 협소건물, 비꼴로	318 page

Extension-Renovation

Mrs. FAN'S PLUGIN HOUSE ǀ 천창을 통해 빛으로 가득 채운 판씨네 플러그인 하우스	206 page
NARVARTE TERRACE ǀ 옥상공간을 활용한 소형주거, 나바르테 테라스	246 page

협소주택 2

41m² - 55m²

LITTLE HOUSE WITH A BIG TERRACE | 큰 테라스를 가진 작은 집	030 page
KAMIUMA HOUSE | 삼각형 대지 위 보이드를 통해 연결방식을 조절한 카미우마 하우스	044 page
SLIDE | 작지만 가족 구성원을 위한 다양한 공간을 담은 주택	056 page
SAIGON HOUSE | 사이공 지역의 특징을 담은 주택	066 page
QUARTER HOUSE | 스윗홈을 담은 쿼터하우스	076 page
YAMASHINA HOUSE | 좁은 대지 위의 공간감 있는 집, 야마시나 하우스	090 page
DENGSHIKOU HUTONG RESIDENCE | L자형 대지 위의 주거 리노베이션, 후통 레지던스	098 page
LITTLE HOUSE, BIG CITY | 도시 속 주거의 지속가능성을 높인 협소주택	108 page
CEDAR HOUSE, PINE HOUSE | 모듈식 삼나무집 · 소나무집	120 page
HOUSE IN THE ORCHARD | 과수원 속의 집	128 page
THE DOVECOTE | 낡은 비둘기장을 개조해 만든 트리하우스	140 page

55m² - 65m²

YEONHUI-DONG 114 | 동네에 자연스레 앉혀진 연희동 일일사	156 page
YANA HOUSE | 밀집된 동네 속 여유로운 박공지붕 주택, 야나 하우스	168 page
PLAYFUL ATTIC HOUSE | 떠있는 다락이 있는 주택 (하우현 성당 옆 주택)	182 page
HOUSE OF TRACE | 과거와 현재가 공존하는 기록의 집	198 page
TOMI HOUSE | 지역날씨에 순응한 주택, 토미 하우스	210 page
KUTTE HOUSE | 실내외 공간 사이의 새로운 가능성을 탐구한 꾸떼 하우스	220 page
CURTAIN COTTAGE | 빅토리아식 주택을 개조해 만든 주거, 커튼 코티지	236 page
POLY HOUSE | 지속가능한 주택 폴리 하우스	248 page
DORIM-DONG MULTI-DWELLING HOUSE 1+2 | 도림동 다가구 주택	260 page
OYAMADAI HOUSE | 대도시를 향해 열린 집, 오야마다이 하우스	268 page

65m² -

NAMHAEJIB | 마당을 품은 상가주택, 남해집	288 page
SHOEI HOUSE | 공간 내외부에서 자연을 느낄 수 있는 주택	300 page
SOYUJAE | 두 자매 가족을 위한 다가구 주택 소유재	310 page
SANDAOSA HOUSE | 두 개의 정육면체가 결합된 산다오사 하우스	318 page
HOUSE IN WAKABAYASHI | 이웃과의 소통을 위한 테라스가 있는 와카바야시 하우스	328 page

CONTENTS_BY SIZE

Narrow-Urban

SLIDE ǀ 작지만 가족 구성원을 위한 다양한 공간을 담은 주택	056 page
SAIGON HOUSE ǀ 사이공 지역의 특징을 담은 주택	066 page
LITTLE HOUSE, BIG CITY ǀ 도시 속 주거의 지속가능성을 높인 협소주택	108 page
POLY HOUSE ǀ 지속가능한 주택 폴리 하우스	248 page
SHOEI HOUSE ǀ 공간 내외부에서 자연을 느낄 수 있는 주택	300 page

Angled-Corner

KAMIUMA HOUSE ǀ 삼각형 대지 위 보이드를 통해 연결방식을 조절한 카미우마 하우스	044 page
DENGSHIKOU HUTONG RESIDENCE ǀ L자형 대지 위의 주거 리노베이션, 후통 레지던스	098 page
DORIM-DONG MULTI-DWELLING HOUSE1+2 ǀ 도림동 다가구 주택	260 page

Typical-Small

LITTLE HOUSE WITH A BIG TERRACE ǀ 큰 테라스를 가진 작은 집	030 page
YAMASHINA HOUSE ǀ 좁은 대지 위의 공간감 있는 집, 야마시나 하우스	090 page
YEONHUI-DONG 114 ǀ 동네에 자연스레 앉혀진 연희동 일일사	156 page
YANA HOUSE ǀ 밀집된 동네 속 여유로운 박공지붕 주택, 야나 하우스	168 page
TOMI HOUSE ǀ 지역날씨에 순응한 주택, 토미 하우스	210 page
KUTTE HOUSE ǀ 실내외 공간 사이의 새로운 가능성을 탐구한 꾸떼 하우스	220 page
OYAMADAI HOUSE ǀ 대도시를 향해 열린 집, 오야마다이 하우스	268 page
SANDAOSA HOUSE ǀ 두 개의 정육면체가 결합된 산다오사 하우스	318 page
HOUSE IN WAKABAYASHI ǀ 이웃과의 소통을 위한 테라스가 있는 와카바야시 하우스	328 page

Compact-Nature

CEDAR HOUSE, PINE HOUSE ǀ 모듈식 삼나무집·소나무집	120 page
HOUSE IN THE ORCHARD ǀ 과수원 속의 집	128 page

Multi-use

QUARTER HOUSE ǀ 스윗홈을 담은 쿼터하우스	076 page
PLAYFUL ATTIC HOUSE ǀ 떠있는 다락이 있는 주택 (하우현 성당 옆 주택)	182 page
NAMHAEJIB ǀ 마당을 품은 상가주택, 남해집	288 page
SOYUJAE ǀ 두 자매 가족을 위한 다가구 주택 소유재	310 page

Extension-Renovation

THE DOVECOTE ǀ 낡은 비둘기장을 개조해 만든 트리하우스	140 page
HOUSE OF TRACE ǀ 과거와 현재가 공존하는 기록의 집	198 page
CURTAIN COTTAGE ǀ 빅토리아식 주택을 개조해 만든 주거, 커튼 코티지	236 page

협소주택

들어가는 글

정부가 내놓는 갖가지 주거정책에도 천정부지로 오르는 집값이 증명하는 현실은, 언젠가 '정원이 있는 나만의 집'을 가질 수 있을 것이라는 꿈마저 희미하게 만들어버린다. 불투명한 미래 속 내 집 마련에 대한 부담감을 가중시키기 보다는, 차라리 지금! 아파트 전세값으로 작더라도 나만의 주거를 갖고자, 작은 자투리땅을 찾아 소형 주거를 건축하는 이들이 많아졌다.

우리는 이 소형주택들을 협소주택이라 부른다. 일본에서 시작한 협소주택의 개념은 이제 국내에서도 흔히 볼 수 있는 주거유형으로 자리잡았다. 그러나 도심 속에 방치되었던 자투리땅이나 소규모 대지에 대한 활용 방안으로 떠올랐던 일본의 협소주택에 비해 우리나라는 그 규모나 성격적인 면에서 정의가 명확히 내려져 있지 않다.

보다 다양한 소형 주거의 형태를 살펴보고자, 이 책에서는 기존의 일반 주택보다는 작은 건축면적을 가진 주택을 다루면서 일반 주택을 포함하여, 쉐어하우스, 전원주택, 세컨드하우스, 이동식 주거 등을 모두 포함하였다. 본 책에 수집된 주거프로젝트의 규모는 건축면적 15m^{2}이상 - 90m^2 미만으로 한다.

이 책은 협소주택이 어떻게 그 협소함을 해결하고 공간을 최대화 하였는지, 각기 다른 사용자의 라이프스타일을 어떠한 건축적 요소를 통해 만들어내고, 주거의 질을 높였는지를 확인하는 것을 목표로 한다. 이를 위해 각 프로젝트에 담긴 물리적·건축적 특성을 분석하여 키워드를 뽑았고(각 프로젝트 개요 상단에 키워드를 표기했다), 그 중 가장 많이 사용된 8가지 건축적 요소들을 선별하여 어떤 형태의 공간으로 나타났는지 정리하였다. 아래의 세 가지 키워드는 주로 협소주택을 설계하는데 있어 건축가들이 중요하게 생각한 속성이다. 그리고 이를 실현시키기 위해 가장 많이 사용한 요소들이 그 밑의 8가지 건축요소들이다.

수납, 수직성, 빛

———

테라스, 보이드 & 중정, 다락 & 로프트, 다양한 층고, 수납, 스킵플로어, 창의 활용, 천창

일반 건축설계에서도 흔히 볼 수 있는 건축적 요소들이지만, 이러한 요소들이 협소한 공간에서 어떻게 다양한 형태의 공간으로 나타났는지 확인해볼 수 있을 것이다.

맹목적으로 큰집을 선호하고 집을 부동산 소유가치로만 평가하던 시대는 가고, 하나하나 자신의 삶의 패턴에 맞춰 디자인된 작은 집에 대한 선호가 점차 확산되고 있다. 이에 따라 건축가들은 작은 집이라고 해서 설계 비용 또는 설계 과정이 규모만큼 적게 소요되지 않는다는 것을 강조한다. 오히려 일반 주거보다 더 까다롭고 어려운 설계와 시공과정을 거치며, 이에 따른 여러 가지 장단점에 대한 내용을 건축가는 물론 건축주도 면밀히 알고 있어야 할 것이다.

자신만의 '작지만 실용적인 소규모 주택'을 짓고자 용기낸 건축주들과 그들의 두드림에 답하고자 수많은 시간을 고뇌하여 설계할 건축가들에게, 본 책이 다양한 협소주택의 형태를 제시하고 비교분석해 볼 수 있는 매개체로 그 역할을 다하길 바란다. 마지막으로 각 프로젝트의 상세한 자료와 인터뷰를 통해 훌륭한 아이디어를 공유해준 참여 건축가분들께 진심으로 감사의 말씀을 전한다.

편집자주

Foreword

The reality that is proved by skyrocketing house prices despite the various housing policies of the government even makes the dreams of "owning my house with a garden" sometime vague. The number of people who seek a piece of land to build a small house with apartment deposit money is now increasing. They want to possess their own homes despite them being small, instead of being increasingly burdened by an oppressive feeling of preparing for their own house in an uncertain future.

We call these small-sized houses on a piece of land "small house." The concept of a small house, which started in Japan, has now become a residence type that is common in Korea. Compared to Japan's small houses, which emerged as a method of utilizing pieces of land or small sites that had been abandoned in downtown areas, those in Korea have not been clearly defined in terms of size or characteristics. To look into more various types of residences, this book includes all houses with a small size such as share houses, country houses, pastoral houses, second houses, and mobile houses, as well as general houses. The scale of the residential projects collected in this book is 00m^2 or less for building area and less than 00m^2 for floor space.

The purpose of this book is to see how small houses maximize space by solving the problem of narrowness, how the different lifestyles of each user were created with what architectural element, and how the quality of the residences was raised. To this end, some keywords were extracted with the analysis of the physical and architectural characteristics of each project, and the words used most among them were selected. The book has organized the type of space those architectural elements are expressed as. The three keywords below are the attributes that architects thought to be the most important in designing small houses. And below them are 8 architectural elements that were used the most to realize these.

Storage, Verticality, Light

terrace, void & courtyard, attic & loft, various heights, storage, skip floor, various windows, toplight

These are the architectural factors that are often seen in general architectural designs. However, we can see how these small ideas are realized as various forms of space.

The days of blindly preferring big houses and only evaluating houses as the value of real estate ownership are in the past. Now, more and more people like small houses that are well-designed according to their life patterns. However, architects emphasize that the cost and process of design is never low and simple because a house is small. The design and construction processes are more complicated than those for general houses. Therefore, building owners as well as architects should thoroughly know the various pros and cons of small houses. In response to the inquiries of building owners who took their courage to build a "small but strong house," I would like this book to provide various types of small houses and play the role of a means of comparison and analysis for the architects who design small houses by agonizing a long time.

Lastly, I would like to thank all the architects who have shared great ideas for us by providing detailed materials and interview for each project.

→ II-156p, Yeonhui-dong 114

협소주택 설계를 위한 조언

건축사사무소 더함, 조한준 소장
ThePlus Architects, architect Cho Hanjun

사람들마다 제각기 삶의 방식과 라이프 스타일이 다르고 다른 용도의 건축보다 가장 사적인 영역이 강조되는 단독주택이기 때문에, 작은 자투리땅에 짓는 협소주택의 경우 건축가는 풀어야 할 숙제들이 너무 많다.

건축가들이 가장 많이 하는 실수 중 하나가 건축가가 당연하다고 생각하는 것이 일반인 건축주에게는 그렇지 않음을 간과하는 것이다. 예를 들어 도면에 그려진 땅의 크기와 모양에서 대해서 물리적인 체감을 달리 한다는 것이고, 도면에 기입된 치수들에 대한 체감도 달리한다는 것이다. 그래서 프로젝트에서 가장 처음으로 했던 것은 건축주에게 현실을 직시할 수 있도록 그 체감되는 크기와 눈높이를 맞추는 일이었다.
기본적인 평면 구상의 스케치를 보여 드린 순간부터는 건축주가 뭔가를 고민할 수 있고 그려 넣을 수 있는 빈 도화지를 주는 순간이다. 아이러니하게도 처음으로 뭔가를 채워 넣어준 것이 결국은 생각을 할 수 있는 기준이 됨과 동시에 빈 도화지가 되는 것이다.
또 하나 건축의 과정은 여러 제품을 비교하여 구매하는 완제품이 아니기 때문에 건축주는 이상적인 100%를 원한다는 것이다. 건축가도 이해 동의하여 끝까지 100%를 쫓기 위해 노력하지만 어느 시점에서는 몇 개 정도를 포기해야 하는 경우가 발생한다. 이럴 때 낙담하는 건축주에게 건축가는 이렇게 말한다. 그동안 작업해왔던 건축의 모든 과정이 매순간 덜 나쁜 것을 선택하는 과정의 연속이었다고 달래준다. 이 말에는 예산문제로 인한 차선택, 공사여건에 따른 불가피한 변경사항, 도심지 공사에서 발생하는 민원문제로 인한 변경 등이 있다.

그리고 협소주택을 지으면서 가장 어려운 점은 충분한 공간을 만들기가 어려워 지하층을 계획해야 할 경우이다. 이 부분은 공사비 상승에 있어서 아주 많은 부분을 차지하는 부분이다. 정말 작은 땅에 지하층을 만들기 위해서는 작은 단위의 치수조차 손해 볼 수 없기 때문에 시공자나 건축가나 많은 노력을 기울여야 한다. 더욱이 지반 상태에 따라 그 공사의 난이도는 천차만별이다. 특별한 경우가 아니면 지하층은 만들지 말라고 당부하고 싶다. 많은 예비 건축주들이 지하층을 계획을 아주 단순하게 생각하는데 이는 큰 오산이다.

그리고 설계기간을 충분히 두고 계획하라고 말하고 싶다. 작은 공간일수록 입체적인 계획이 무엇보다 중요하다. 내부의 충분한 공간감과 효율적인 공간 구성을 위해서는 더 많은 고민을 해야 한다. 경험적으로 보았을 때 설계가 어려운 경우 공사는 몇 배 더 어렵다는 것이다.
협소주택은 거의 도심지 내에 지어지는 공사이기 때문에 공사가 어렵다. 민원에서 자유로울 수가 없고, 도로에 지나가는 행인이나 차량으로 인해 공사차량의 진입이 어렵다. 공사기간이 예정처럼 진행되지 않는 것은 부지기수다. 만일 정해진 시간 내에 설계를 하고 예정 시간 내에 준공을 하여 이사를 하는 일정을 잡고 있는 예비 건축주가 있다면 다시 한번 시간에 여유를 두고 계획하라고 알려주고 싶다.

많은 협소주택 예비 건축주에게 말하고 싶은 것은 협소주택은 결코 싸게 지을 수 있는 집이 아니라는 것이다. 오랜 기간 예산을 책정하는 방법 중에 하나인 평당 공사비는 협소주택에 있어서는 무용지물이다. 주택의 크기의 문제가 아니기 때문에 경험 있는 건축가와 잘 상의해야 할 것이다.
공사과정을 보면 현장에서 작업자들의 고생은 이루 말할 수 없다. 골조공사가 진행이 될 때에는 형틀목수와 철근 배근공, 전기배선 작업자 등 좁은 공간 내에 10명 정도가 동시에 어깨를 부딪혀 가며 일해야 하는 경우가 부지기수다. 건물이 완성될 이미지와 도면만 볼 것이 아니라 그 과정에 대한 고려를 염두에 두어야 한다.

마지막으로 건축주와 시공자와 건축가는 파트너가 되어야 한다. 대가를 지불하고 그에 대한 결과만을 바라는 구매자로서의 건축주가 아니어야 하고, 돈을 받았으니 기술력과 노동을 제공하는 업자로서의 시공자가 아니어야 하고, 건축 과정에서 그림과 도면을 제공한 단순한 설계자로서의 건축가가 아니어야 한다. 앞서 말한 내용대로 도심지 내 어려운 공사이며 많은 변수들이 도사리고 있기 때문에 항상 같이 의논하고 그 해결책에 대한 고민을 공유하며 서로를 다독여야 하기 때문이다. 이 부분은 협소주택에만 해당 되는 사항이 아니지만 집을 짓기를 희망하는 예비 건축주들에게 부탁하고 싶은 사항이기도 하다.

As people have their own life styles and private areas are emphasized more in detached houses than buildings with other purposes, architects face too many problems to solve in the case of a small house built on a piece of land.

One of the common mistakes architects make is to overlook the fact that what they take for granted as architects is not what building owners think about as ordinary people. For instance, architects and building owners see the size and shape of land on drawings differently. Their physical feelings about the measurements on drawings are also different. That is why the first thing I did for this

→ I-170p, [CREVICE] 1740

Advice For Planning Small Houses

project was to adjust the expectation level of the building owner to be the same as mine in order for them to face up to the reality.

The moment I showed a basic two-dimensional design to the building owner was the moment I gave a blank drawing paper to the building owner for them to think about something seriously and draw or put something on the paper. Ironically, conveying a drawing, in which something is shown, becomes a standard for thinking and blank drawing paper to the building owner.

Another difficulty is that building owners usually want a one hundred percent ideal building even though the construction process is not a complete product and is purchased after the comparison of several products. Architects also agree with building owners and make efforts to obtain one hundred percent satisfaction until the end. However, at some point, they cannot help giving up a few things. At such a time, architects talk to the disappointed building owners like this. The entire process of construction is the continuity of moments to select less bad things. Architects assure building owners like that. What "less bad thing" means includes taking the next best thing due to budget problems, inevitable changes caused by construction conditions, or changes caused by civil complaints in downtown construction.

And the most difficult thing in building a small house is planning the basement floor, because there isn't sufficient space. This contributes a lot to increase construction costs. To make a basement floor on a really small lot, builders and architects should make a lot of effort not to suffer a loss even with the smallest measurements. Moreover, the difficulty level of construction varies depending on the state of the ground. I want to ask not to make a basement floor unless it is a special case. Many future building owners think the plan of a basement floor is simple, but this is a misjudgment.

I also want to tell them to plan with a sufficient design period. A three-dimensional plan is more important than any other thing for a small space. More thinking is needed for a sufficient feeling of the

space inside and efficient space composition. From my experience, construction is a few times more difficult when the design is difficult. Construction can't be free from civil complaints, and construction vehicles have difficulty entering because of pedestrians or passing cars on the road. It's very common for the construction time to be different from the plan. If there are future building owners who have scheduled the design of a building within a particular time frame and want to complete construction and move into the house in the expected time, I want tell them to plan again and take their time.

What I really want to tell many future building owners is that a small house cannot be built at a low price. The construction cost by pyeong, one of the old budgeting processes, is a useless thing when you build a small house. As this is not the problem of house size, building owners should consult well with architects.

The hard labor of workers in the construction process is beyond description. During frame construction, about 10 workers such as concrete form carpenters, rebar reinforcement engineers, and electric wiring workers should work together in a small space with their shoulders bumping into each other. Instead of looking at images of a completed building and drawings, the process should be kept in mind.

Lastly, building owners, builders, and architects should become partners. The building owners should not be just purchasers who only want the results of their payment, while builders should not be just dealers who provide technology and labor in return for money. And architects should not be just designers who only provide images and drawings in the process of construction. As I mentioned before, small houses usually require a difficult construction process in the downtown area where lots of variables lurk. That is why the three parties should always discuss things together, share their thinking about solutions, and soothe each other. This is what I also want to ask for all future building owners who want to build a house, not only for the owners of small houses.

주변환경을 포용하고 디자인하는 주거건축
Housing Architecture Designing and Embracing the Environment

Frontofficetokyo

우리는 도시라는 공간을 거시적 관점에서 보는 것에 관심을 두고 있으며, 건물을 설계할 때 주변과 이웃을 고려하려고 노력한다. 도쿄처럼 인구 밀도가 높고 빼곡한 도시에서는 직관적이지 않은 생각처럼 들릴 수 있겠으나, 건물이 사생활 때문에 주변 장소와 분리되기보다는 그곳을 포용해야 한다고 생각한다. 순수하게 재정적 관점에서 보면, 도쿄의 땅값이 매우 높기 때문에 보유한 땅을 건축물 설계에 모두 사용하지 않는다면 비용 측면에서 낭비이다. 게다가 우리는 도시에서 개방적인 주택을 지으면 바로 옆 건물과 시야 거리가 짧을 때에도 집이 넓어 보이는 느낌을 준다는 것을 알고 있다. 항상 유리 건물을 지어야 한다는 뜻이 아니라, 독특한 건물 이상의 것을 만들어야 한다는 뜻이다. 특히 우리가 지을 주택에서는 이 점을 고려하여 건물 전체의 주변 풍경을 디자인하려고 노력했다. 나무를 가득 심거나 전체를 콘크리트로 만들 수도 있지만, 풍경으로 주택이 지닌 시각적 한계를 넘어선다면 집이 더욱 편안하게 느껴질 것이다.

일본 건축의 관행은 권위적이기보다 다양한 변수를 반영하기 때문에 흥미롭다. 우리는 주어진 부지 위에서 건물주에 최대한 맞춘 주택을 여러 가지 방식으로 건축할 수 있다. 이는 건축가로서 일종의 호사지만, 여의치 않은 상황에서도 좁은 부지에 맞는 훌륭한 설계를 찾는 유일한 방법은 열린 마음으로 다가가고, 형태나 소재를 우선순위로 정하지 않고 모든 가능성을 모색하는 것이라 생각한다. 우리의 경우에는 실물이나 디지털로 마을 모형을 만들고 설계 과정에서 여러 가지 옵션을 상황에 맞게 테스트한다. 시도하는 모든 변수 간에 연관성이 있다면, 공간을 연결하고 기존 풍경에서 기회를 찾아 활용할 수 있다.

→ II-268p, Oyamadai House

We are certainly interested in a zoomed-out view of the city, and try to consider the neighborhoods as much as the buildings we are designing. It is perhaps counter-intuitive in a city as dense and tightly packed as Tokyo, but we feel that buildings should embrace the location as much as possible rather than retreat from it for the sake of privacy.

In purely financial terms the cost of land is so high that any part of the property that is not incorporated into the design is a costly waste. At the same time we have learned that opening up to the city makes a home feel larger, even when the view is as short as the building next door. This does not mean that we should always make a glass building, only that we should go beyond making a singular object. With this in mind, we make an effort to design the landscape around all of our buildings, especially in the case of our houses. The landscape can be filled with trees or made entirely from concrete, but if it extends the visual threshold of a home then it will be more comfortable.

The Japanese building code is particularly interesting because it is parametric rather than prescriptive so we are able to build any number of ways on a given site, matching the house to the owner as much as possible. This is a kind of luxury, but even if it were not the case, our feeling is that the only way to find a good design on a small piece of land is to approach it with an open mind, and to explore possibilities without any prior commitment to form or material. In our case this means we begin by making a physical or digital model of the neighborhood and testing out a range of options with the context as part of the design process. If there is a connection between all of the variations that that we try out, it is to connect spaces, and to find opportunities in the existing landscape that we can take advantage of.

Small house in japan

Hiroto Suzuki architects and associates

단독주택은 도심 지역에서 거주하는 핵가족에게 적합하지 않다. 일반적으로 낮 시간 동안 부모는 직장에 가고 자녀들은 학교에 가기 때문에, 핵가족은 아침 식사를 할 때나 잠을 잘 때와 같이 아주 적은 시간만을 집에서 보낸다. 또한 거의 대부분의 자녀들이 고등학교를 졸업하고 나면 독립하며, 시간이 지나면서 가족 구성원 수가 바뀌게 된다.

나가야(Nagaya)는 예전에 도심 지역에서 유행했던 협소주택 스타일의 집이다. 화장실과 욕실이 한 공간에 있으며, 거실은 소파 베드가 있는 침실로도 사용된다. 자기만의 방이 있는 사람은 집주인뿐이다.

최근 서구화의 영향으로 단독주택에 가족 구성원별로 각자의 방을 만드는 것이 인기를 끌기 시작했는데, 그 중 일부는 다소 과한 감이 있다. 그래서 도심 지역의 협소주택 설계 시 일본인 가족에게 불필요한 것은 없애버렸.

도심 지역에는 수많은 상점과 레스토랑, 그리고 방문객들이 머물 수 있는 호텔이 있으므로, 살면서 필요한 것들은 주변 환경에서 해결할 수 있다. 굳이 거주지에 여분의 공간을 둘 필요가 없는 것이다.

가장 중요한 것은 거실을 식사를 하고, 대화를 나누고, TV를 보고, 독서를 하고, 숙제를 하고, 취미 생활을 즐기기에 알맞은 편안하고 실용적인 공간으로 만드는 것이다. 몇 가지 기능을 거실로 통합시키는 것은 협소주택 설계 방식 중 하나다.

Single family dwelling is not suitable for nuclear family living in urban area. Parents go to work generally and children go to school during a day thus nuclear family spent a little times in their house, just for eating breakfast and sleeping.

Almost all children become independent after finish high school. Number of family member is changing with the passage of time.

Nagaya is very compact house style that had been popular dwelling in urban area old days. Washroom and bath room, well are held in common, and living room was used as bedroom with futon. Just only master has own private room.

Recently, under the influence of westernization, it have popularized that single family dwelling have each family member's private rooms and bathroom, but some of these excessive. I took off what is unnecessary for Japanese family when designing small house in urban area.

Urban area has lot of shops, restaurant and hotels for guest, we can rely on surroundings for living, no need extra space for living.

Most important thing is make living room comfortable and effectively utility for taking a meal, enjoy chatting, watching TV, reading book, doing homework, hobby and so on. Integrating some function into living room is one of method for planning small house.

→ II-328p, House in Wakabayashi

작은 면적, 넓은 공간을 실현하기 위한 8가지 건축요소

천창

테라스

창의 활용

스킵플로어

수납(&맞춤가구)

다양한 층고

다락 & 로프트

보이드 & 중정

toplight

terrace

various window

skip floor

storage (& customized furniture)

various story height

attic & loft

void & courtyard

8 elements for small house

천창
toplight

밀도가 높은 도심지역의 주택들은 프라이버시 등의 문제로 인해 벽에 창을 내어 자연채광을 확보하기가 쉽지 않다. 이런 경우에 효과적으로 채광을 내부에 들일 수 있는 방법 중 하나가 천창이다. 천창으로부터 유입되는 자연광으로 인해 실내의 밝기가 유지되고, 하늘의 풍경까지 감상할 수 있는 개방감을 준다.

It is not easy to let in natural light by putting windows in a wall in houses in a high-density urban areas because of issues such as privacy. In this case, an effective way to bring in light is with a toplight. The natural light coming in through a toplight maintains the brightness indoors and provides openness since the sky can be appreciated through it.

→ I-66p, PH Lavalleja

→ I-66p, PH Lavalleja

주택 밀집지역에서 지붕에 천창을 내면 실내로 유입되는 광량을 충분히 확보할 수 있다. 두 공간을 연결하는 계단 위쪽의 천창은 공간 선체를 밝히는 자연조명 역할을 하고 개방감을 준다.
In a high-density housing area, sufficient light can be secured by putting a toplight in the ceiling. A toplight above a staircase connecting two spaces brings in natural light that brightens the whole space and provides openness.

→ II-248p, Poly house

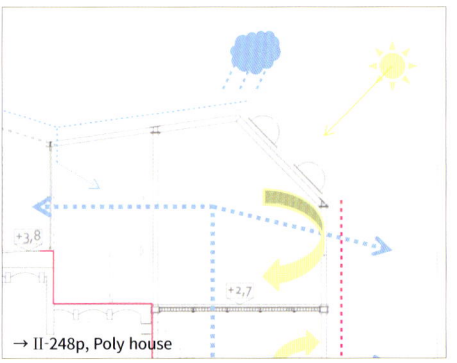
→ II-248p, Poly house

큰 창문이나 천창 등 여러 곳의 창문으로부터 다각도로 들어오는 자연광이 만들어낸 빛의 변화가 공간에 생동감을 더한다.
Changes in natural light coming from various angles through the windows in various places such as big windows or toplights adds vividness.

toplight

→ I-160p, Nest

→ I-56p, House at Hommachi

윗층 천창에서 들어오는 채광은 계단실 전체를 밝히고 계단실을 통해 아래층 공간까지 밝게 한다.
The light from a toplight in the ceiling brightens the entire staircase and even the interior downstairs by coming in through the staircase.

주변 건물로 둘러싸인 환경에서 주거인의 생활편의를 위해 주방 공간 위에 북측 지붕에 천창을 만들었다. 자연광이 주택 깊은 공간까지 밝혀주며 공기를 위아래로 순환시켜 건물 전체를 환기시켜 준다.
In an environment surrounded neighboring buildings, if a toplight is placed in the northern ceiling above the kitchen for the convenience of the residents, natural light brightens the space deep in the house and it ventilates the whole building by circulating air up and down.

→ I-100p, House in Kawasaki

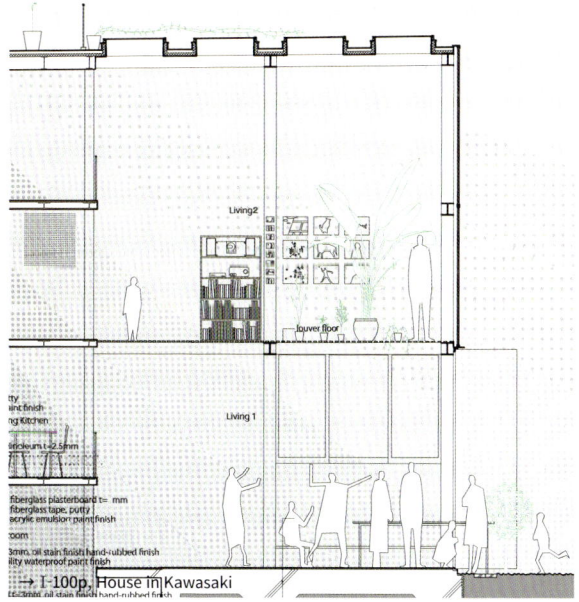
→ I-100p, House in Kawasaki

2층 천창에서 들어오는 자연광은 일부 유리바닥 또는 루버바닥을 통해 1층 공간까지 밝게 만든다.
The natural light flowing in through a toplight on the second floor even partially brightens the first floor through the glass floor.

테라스
Terrace

테라스나 발코니 또는 옥상 등의 외부공간을 내부공간으로 끌어들여, 좁은 내부를 확장해서 사용할 수 있다. 또한, 채광이나 통풍 등 자연과 가까운 주거의 삶을 즐길 수 있는 여유와 쾌적함을 주고, 잘 꾸며놓은 외부공간은 집 안에서의 다양한 풍경을 선사한다.

By drawing outdoor spaces such as the terrace, balcony and rooftop indoors, a small interior area can be expanded. These spaces provide leisure and comfort close to nature with light and wind, and a well-decorated outdoor space presents a variety of landscapes inside the house.

LDK와 연결된 테라스

→ II-30p, Little house with a big terrace

LDK와 이어지게 테라스를 만들면, 테라스의 문을 완전히 열어 두었을 때 실내외 구분이 모호해져 연결감이 형성되어 공간이 더욱 넓어 보이게 한다.

If a terrace is built to be connected with the LDK, the connection between indoor and outdoor spaces is formed when the door to the terrace is fully opened and the boundary between indoor and outdoor spaces becomes blurred and makes the space look large.

→ II-236p, Curtain Cottage

옥상 테라스

→ I-30p, Rooftop house

→ II-44p, Kamiuma house

옥상은 외부와의 시선이 차단되는 공간이므로 테라스나 발코니를 마련하기 어려운 경우에, 프라이빗 가든으로 적용하여 도시의 풍경을 감상하는 다목적 공간으로 사용할 수 있다.

The rooftop allows for privacy by blocking sight from outside but offers natural light inwards. It can be used as a multi-purpose space to enjoy the scenery of the city.

terrace

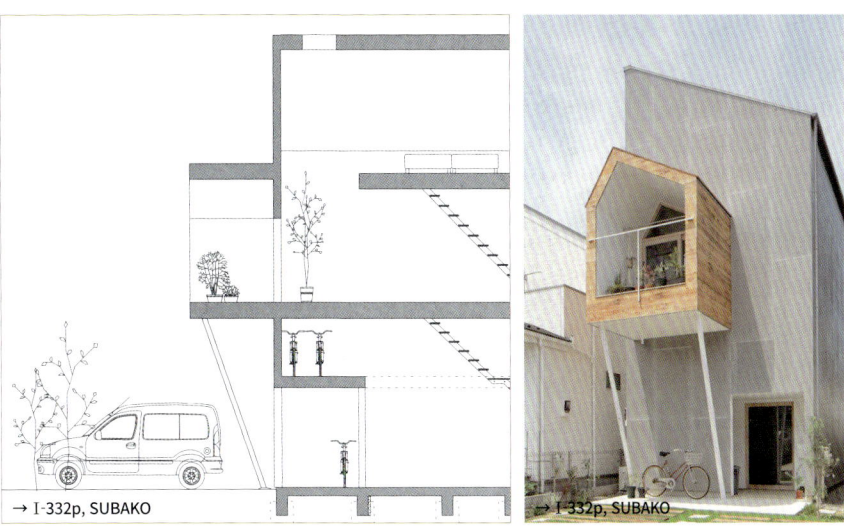

→ I-332p, SUBAKO

발코니

지상의 테라스를 만들기 어려운 경우, 발코니를 설치하여 외부와 내부 사이의 중간영역을 만들고 주고, 건물 입면에 다양성을 주는 효과를 가져올 수 있다.
When it is difficult to make a ground level terrace, installing a balcony makes a middle area between the inside and outside spaces and gives variety to the façade of the building.

→ I-100p, House in Kawasaki

→ I-246p, Narvarte Terrace

슬라이딩도어 & 폴딩도어

미닫이 형식의 풀 오픈 도어 또는 폴딩도어를 설치하면 내외부를 잇는 중간 영역이 보다 자연스럽게 형성되고, 필요에 따라 쉽게 공간을 연결·분리할 수 있는 장점이 있다.
외부공간을 최대한 내부로 끌어들이고자 벽면 전체를 개구부로 꾸미는 경우가 있는데 이때 슬라이딩도어나 폴딩도어가 유용하다.
If a fully opening sliding type door or folding door is installed, the area between the inside and outside spaces is formed more naturally and the space can be easily separated or connected when the occasion demands. The whole wall is sometimes constructed as an opening to draw the outside space in as much as possible. At this time, a sliding or folding door is useful.

8 elements for small house

창의 활용
various window

창은 빛, 바람 그리고 조망을 확보하는 역할을 한다.
특히 협소한 공간일 경우 다양한 창의 활용에 따라 좁은 공간을 넓게 보이는 효과를 얻는 동시에 자연 환기와 채광을 가능하게 한다.
또한, 주거 밀집 지역일 경우는 측창을 활용하여 통풍과 채광 동시에 프라이버시를 확보할 수 있다.

Windows secure light, block wind and provide a view. In small spaces in particular, the various uses of windows can make a narrow space look spacious and enable natural ventilation and lighting. In the case of a high-density residential area, side windows can be used for ventilation and lighting as well as privacy.

바람이 드나드는 통로

→ I-30p, Rooftop house

바람이 드나드는 입구와 출구가 되는 위치에 창을 두어 집 전체의 공기순환을 이루게 한다. 쌍여닫이창문을 열어 젖히면 자연 채광과 환기가 이루어지며 야외 공간을 실내로 들여온 듯한 효과까지 얻을 수 있다.

Windows were placed in locations at the entrance and exit to make overall air circulation in the house possible. When a double hung window is opened, natural lighting and ventilation are possible, and it even looks like the outdoor space is inside.

various window

공간에 깊이감을 주는 측창

높은 위치의 측창을 통해 자연광이 위에서 바닥까지 전달되어 이전의 어두웠던 공간을 빛으로 채우는 동시에 좁은 공간도 넓게 보이는 효과를 얻을 수 있다. 또한, 옆집의 큰 창이 있어도 서로 시선이 닿지 않아 프라이버시를 지킬 수 있다.

A side window in a high place can send sunlight to the ground. Then, a dark space is filled with light and a narrow space looks spacious. Also, privacy is protected because even if there is a large window in the next house, people do not see each other.

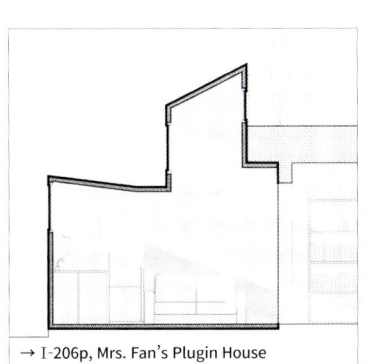
→ I-206p, Mrs. Fan's Plugin House

→ I-206p, Mrs. Fan's Plugin House

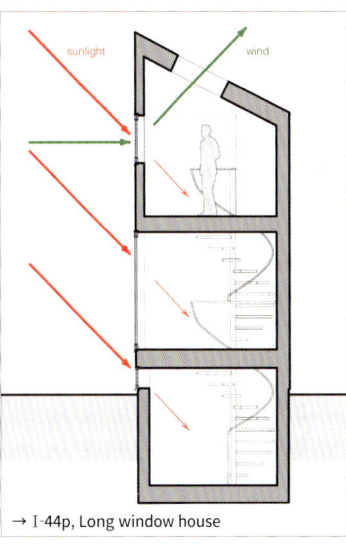
→ I-44p, Long window house

공간을 더 넓게 느껴지도록 만드는 대형창

→ I-160p, Nest

→ I-160p, Nest

큰 창은 외부로 시선이 향하도록 유도해 외부공간을 실내로 끌어들여 시원시원한 공간을 만든다. 두 배로 높은 생활 공간 위의 상단 창문을 통해 햇빛이 충분히 들어오며, 하늘이 보이기 때문에 탁 트인 느낌을 준다. 발코니의 루버는 내부에서의 가시성을 제공하는 한편, 사생활 보호 수준 조절을 위해 외부로부터의 시선은 차단한다.

A large window can make a space look open by drawing the outside space indoors and causing people to look outside. Sufficient sunlight comes in through a window in the upper area of a living space which is twice as high as others. Also, as the sky is seen through it, home dwellers can feel openness. The louvers on the balcony provide visibility from the inside and block the sight from the outside to adjust the privacy protection level.

8 elements for small house

스킵플로어
skip floor

스킵플로어는 협소한 면적에서 공간을 넓어 보이게 하는 요소 중 하나이다. 바닥레벨을 상하로 어긋나게 한 배열은 각 공간 간의 리듬을 만들고, 시선이 교차하면서 공간의 흐름에 연속성이 생겨 시각적으로 공간이 확장되어 보이게 한다.

A skip floor is a factor that makes a small area look spacious. The up and down arrangement of the ground level creates rhythm between each space and gives continuity to spatial flow as sightlines cross, thereby making the space look extended.

대지의 높이차로 인해 발생한 제약조건을 반층씩 높아지는 스킵플로어 구성으로 해결하였다. 낮은 대지쪽에는 반지하층 주차공간을, 높은 대지쪽은 1층 진입 공간으로 만들어 지형을 동선 형성에 이용하였다. 이로인해 시야는 열려있되, 공간적으로는 분리되었다.

A constraint caused by the height differences of the site was solved with the construction of a skip floor. The semi-basement level parking area was made on the side with low land while the entrance to the first floor was made on the high side. In this way, the topography was used for the formation of the circulation. This provides an open view despite spatial separation.

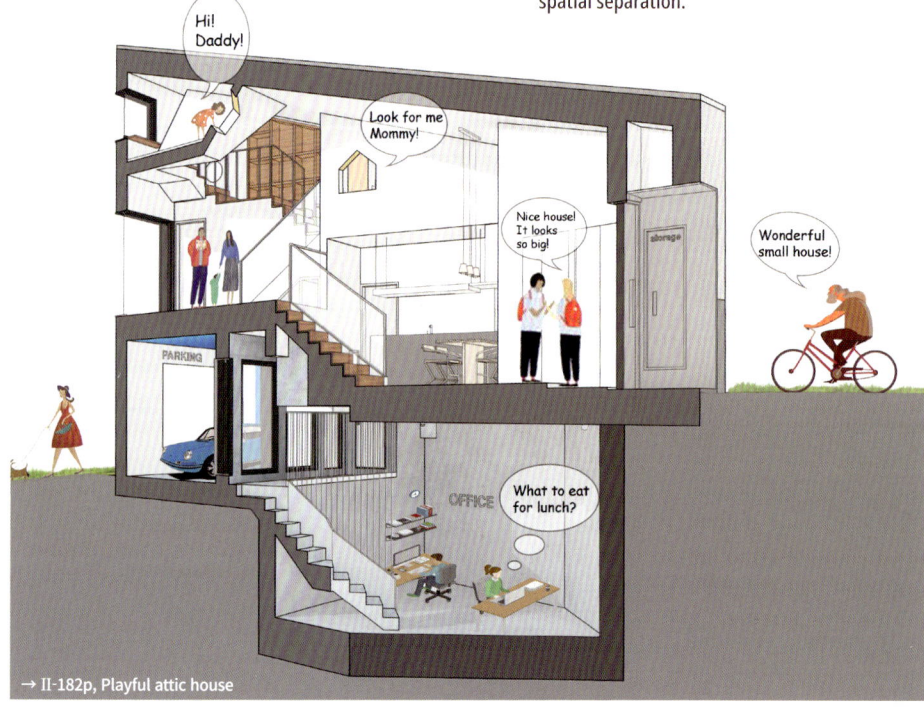

→ II-182p, Playful attic house

skip floor

밀도가 높은 도심의 주택은 수평구조보다는 수직구조로 이루어지기 때문에 주거 내에서 계단이 중요한 요소이다. 이 계단에 의한 사용자의 피로도가 높아지는 만큼 스킵플로어를 잘 사용하면 수직성에서 오는 불편함을 줄일 수 있기도 하다. 또한, 대지 단차에서 오는 문제점을 스킵플로어로도 해결할 수 있다.
A staircase is an important element in a house with a vertical structure in a downtown area with high density, but can increase the fatigue of users. However, the use of a skip floor can reduce the discomfort, and can also solve the problem of differences in land levels.

→ I-258p, Circulate house

→ I-288p, H4912

→ I-272p, S1927

협소주택의 수직배치에서 상하층의 물리적·심리적 거리감을 단축시켜 소통하게 하는 것은 매우 중요한 사안이다. 이때 계단의 경사도를 완만하게 하거나, 반층 정도의 계단참을 마치 스킵플로이치럼 보이도록 연장해 주면 아래위 공간이 보다 자연스럽게 연결된다.
스킵플로어로 인해 내부공간에 생긴 단차는 자연스럽게 영역을 분리하며, 올려다보고 내려다보는 다각도의 시선의 방향을 만든다.
In the vertical placement of small houses, it is critical to reduce the physical and psychological sense of distance between upstairs and downstairs and make communication possible. If the slope of a staircase is gentle or a half-story stair landing is expanded to look like a skip floor, the upstairs and downstairs spaces are naturally connected. The difference in interior levels caused by a skip floor naturally divides the space and creates multiple directions for people to look up and look down.

수납
storage

수납은 꼭 협소주택이 아니더라도 어느 주거형태에서나 필수적으로 적절히 마련되어야 하는 요소이다. 특히 면적이 작은 협소주택에서는 데드스페이스나 맞춤 가구를 이용하여 최대한의 수납공간을 만들어내는 것이 관건이다. 특히, 맞춤가구는 공간과 조화를 이루는 일체감을 주어 보다 정돈된 느낌의 실내분위기를 만들어 낸다.

Storage is a factor that should be necessarily and appropriately prepared for in all residence types. In small houses in particular, whose area is narrow, it is crucial to make maximum storage space by using dead space or customized furniture. Customized furniture in particular creates an interior atmosphere that looks organized by giving a sense of unity where the furniture harmonizes the space.

벽면을 활용한 수납

→ II-90p, Yamashina house

→ II-108p, Little House. Big City

Top, House in Wakabayashi

벽면 수납은 벽의 바닥에서 천장까지 수납에 활용할 수 있기 때문에 넓은 수납공간을 확보할 수 있고 동시에 인테리어 효과도 줄 수 있다.

Wall storage can not only secure a large storage space, because the area from the ground to the ceiling can be used for storage, but can also give an interior effect.

높은 층고를 활용한 천장 수납공간

→ II-268p, Oyamadai house

→ II-268p, Oyamadai house

높은 층고를 활용해 수납공간을 천장에 매달아 넓은 생활공간을 확보하였다. 잠시 보관하는 물건이나 자주 쓰는 물건은 이러한 선반이나 수납장을 이용해 마음껏 보관할 수 있다.

Shelves were hung from the ceiling using the height of the story, and thus a large living space was secured. Goods stored temporarily or used frequently can be stored easily using such shelves or a storage closet.

계단 아래 수납공간

→ II-56p, Slide

벽계단 아래의 수납장은 보통 맞춤가구로 그 실용성이 높다. 계단 밑 수납장을 닫으면 마치 벽면처럼 보여 깔끔함을 선사한다.

storage closet under a staircase usually has high practicality as the customized furniture. When the storage closet under the staircase is closed, it looks neat, like a wall.

storage / various story height

다양한 층고
various story height

천장의 높낮이를 달리함으로써 공간의 획일성과 단조로움을 피하고, 공간감에 다양한 변화를 줄 수 있다.
천장의 높이 변화는 위층의 바닥 단차를 만들어 자연스럽게 공간 프로그램의 분리가 이루어진다.

Different ceiling heights can provide variety to a space instead of conformity and monotony. The variance in ceiling height makes the ground level different, thereby naturally separating the space program.

→ I-206p, Mrs. Fan's Plugin House

천장의 높낮이에 따라 공간의 구획이 생겨 자연스레 분리가 이루어진다. 또한 보다 다양한 입체감과 변화를 느낄 수 있어, 작은 면적 안에서의 단조로움을 피할 수 있다. 천장 높이의 변화 또는 구배는 공간에 완급을 주는 매력을 지닌다.

→ II-288p, Namhaejib

→ II-220p, Kutte house

→ II-220p, Kutte house

The height of a ceiling divides a space and thus causes a natural division. Also, it makes various three-dimensional effects and changes, thereby reducing the monotony in a small area. As such, the variance in ceiling height or ceiling gradient has an attraction that gives variety to space.

아래층 천장의 높낮이 차는 위층 바닥의 단차를 가져온다. 이는 평면적으로나 단면적으로 역동적인 내부 공간의 흐름을 만들어 준다.

The height difference downstairs makes a ground level difference upstairs. It makes a dynamic flow of internal space in terms of both two dimensions and cross section.

다락 & 로프트
attic & loft

다락은 주거공간 내부에서 생기는 자투리 공간을 활용하기에 아주 적합한 요소이다. 다소 폐쇄적인 공간감을 줄 수 있으나, 천창이나 작은 창으로 적절한 개방감을 주면 어린아이들을 위한 놀이방 또는 수납공간, 어른들의 서재로도 사용할 수 있는 등 다목적 공간으로 활용될 수 있다.
An attic is a suitable factor to utilize the spare space generated inside a residential space. It gives a somewhat closed sense of space, but if an appropriate openness is added with a skylight or small window, the attic can be used as multipurpose space such as a play room for children, storage space, or adults' study room.

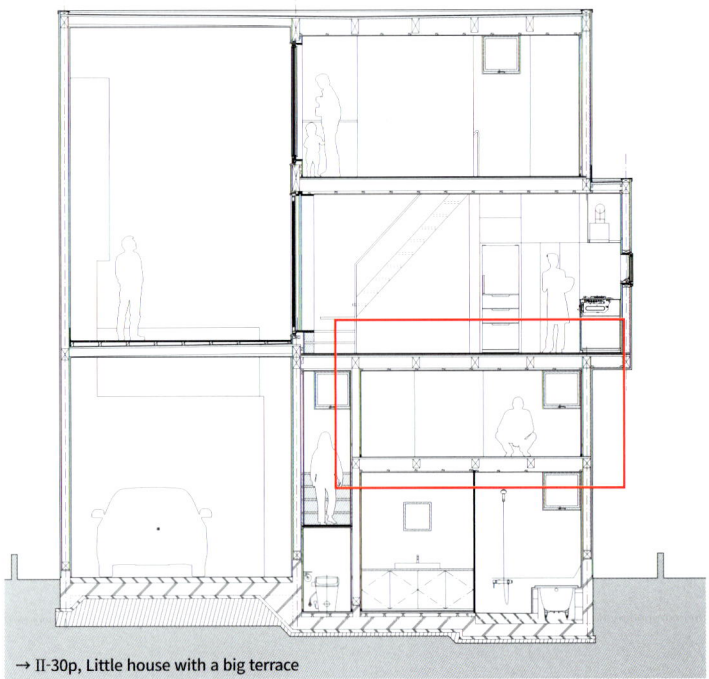

→ II-30p, Little house with a big terrace

대지면적이 한정된 상황에서 테라스와 같은 외부공간을 만드는데 사용한 면적을 만회하기 위해, 1층과 2층 사이에 창고로 사용할 수 있는 천장높이 1.4m 미만의 다락을 만들었다. 이는 실용적 공간을 확보하는 동시에 거실과 테라스를 높여 채광을 개선하는 효과를 가져온다.
To make up for the area used for outdoor space like a terrace in a limited land area, an attic with a ceiling height of less than 1.4 m was built. It can be used as a warehouse in the area between the first floor and second floor. This secures a practical space and improves lighting by raising the living room and terrace.

attic & loft

거실 끝켠에 로프트를 두어 계단참으로 활용할 수 있다. 특히 이 로트프의 연창은 내부에서 자연을 감상하고, 환기도 가능하게 한다. 이러한 로프트는 서재 또는 수납 등 다목적 공간으로 활용 가능하다.

A loft placed in the corner of the living room can be used as a stair landing. In particular, the band window of the loft enables the appreciation of nature indoors, and ventilation. Such lofts can be used as multipurpose spaces such as storage space or study rooms.

→ I-308p, Y house

→ II-182p, Playful attic house

다락은 아이들에게 좋은 놀이터가 될 수 있다. 밑의 공간을 내려다볼 수 있는 내부창과 기울어진 천장, 여러 층과의 연결은 아이에게 흥미로운 공간적 경험을 제공한다. 또한, 추후 아이의 성장에 따라 취미공간, 작은 서재로도 사용 가능하다.

An attic can be a good playground for children. An interior window through which one can look down on the space below, slanted ceiling, and connection of stories provide children with an interesting spatial experience. Also, it can be changed to a hobby room or small study room later as children grow.

→ II-182p, Playful attic house

→ II-182p, Playful attic house

보이드 & 중정
void & courtyard

보이드와 중정은 자연환경, 거주자, 외부 공간을 연결시켜주는 기능을 한다. 주거 밀집 지역에서는 정원을 위한 대지를 따로 마련하기 쉽지 않을 뿐만 아니라 프라이버시 보호의 문제도 있다. 이때 건물 내부에 보이드와 중정을 마련하면 외부의 시선이 차단된 상태에서 자연스레 채광과 환기가 가능하다. 또한, 열린 구조를 통해 좁은 공간이 넓어 보일 수 있다.

Voids and courtyards connect the natural environment, home dwellers, and outside spaces. It is not easy to get land for a yard in high density residential areas, and there is also the issue of privacy. The void and courtyard prepared inside a building enable lighting and ventilation without the need to care about other people's views. Also, an open structure makes a small space look broad.

실내외 요소간의 관계를 연결하는 보이드

→ II-44p, Kamiuma house

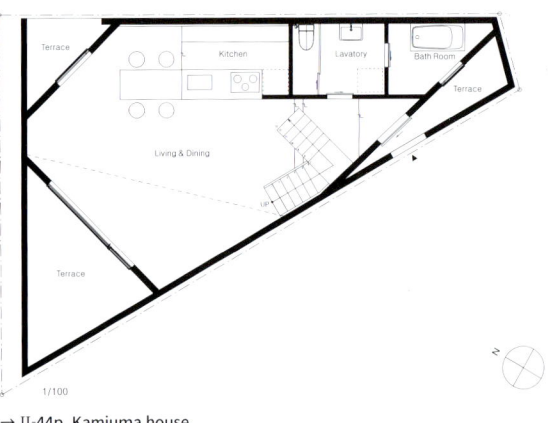

→ II-44p, Kamiuma house

실내의 보이드를 통해 거주자의 시선이 서로 연결되며, 햇살·소리·바람 외 자연환경 요소와도 연결이 이루어진다. 위의 도면에서는 세 군데 보이드가 이러한 실내외 요소 간의 관계를 조절하는 것을 볼 수 있다.
이렇게 연결 방식을 조정함으로써 적당한 수준에서 외부와 상호작용할 수 있고 궁극적으로는 주변 환경과 조화롭게 공존할 수 있다.

The void inside connects the sight of the home dwellers to each other and also to the natural environment such as sunshine, sound, and wind. Three voids adjust such a relationship between indoor and outdoor elements. With this adjustment of the connection, people can interact with the outside at a proper level, and can ultimately coexist harmoniously with the surrounding environment.

void & courtyard

다채로운 공간감을 느낄 수 있는 보이드 연결공간

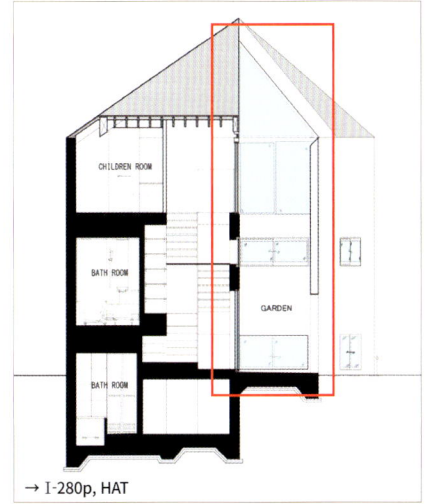
→ I-280p, HAT

→ I-280p, HAT

→ I-140p, Trans

밀도 높은 도심 내 주택에 중정을 계획하면, 외부에서 보기에는 다소 폐쇄적으로 보일지라도 프라이버시를 보호하면서 내부 공간은 자연광이 풍부하며 개방적인 느낌을 준다. 또한, 중정을 통해 연결된 실내를 통해 다채로운 공간감과 거리감을 느낄 수 있다.

A courtyard design in a high density area can protect privacy because it looks closed from outside, but the inside space has abundant sunshine and feels very open. Also, the house interior connected to a courtyard gives various senses of space and distance.

계절감을 선사하는 중정

→ I-152p, H33617

→ I-152p, H33617

건물 내 보이드를 통해 내부에 자연 채광을 끌어들이고 자연 통풍을 원활하게 한다. 또한, 도시 속 일상생활 속에서도 계절을 느낄 수 있는 정서적인 기능을 한다.

The void inside a building draws in natural lighting and makes the natural draft smooth. Also, it has an emotional function that lets home dwellers feel the changing seasons even in busy daily city life.

ROOFTOP HOUSE 30p
LONG WINDOW HOUSE 44p
HOUSE AT HOMMACHI 56p
PH LAVALLEJA 66p
SLICE 78p
MINI HOUSE KODA 84p

15m² - 25m²

23.60m²

Rooftop Terrace Big window

Location
Tokyo, Japan
Use
House
Site Area
59.92m²
Built Area
23.60m²
Total Floor Area
70.38m²

Floor
B1 - 2F
Exterior Finish
Lysin spraying
Interior Finish
Emulsion paint finish

Project Architect
Tsuyoshi Kobayashi
Construction
Show-Yo
Photographer
Koichi Torimura

Rooftop House

옥상공간을 활용하여 도시환경의 기회와 가능성을 담은 집

another APARTMENT

15m² - 25m²

Rooftop house

**옥상 공간을 효과적으로 잘 활용하면 도시 환경이 지닌 기회와 가능성을 극대화할 수 있다.
도심 주거 지역이라는 위치에도 불구하고 목재 데크를 들어올린 듯한 옥상에서는 주변 경관이나 불어오는 바람을 통해 자연과 계절의 변화를 느낄 수 있다.**

도쿄 주택가에서 흔하게 볼 수 있는 이 작은 부지는 60m²로 면적이 좁고 규제가 까다로운 지역에 위치해 있다.
규정에 따라 큰 건물을 지을 수는 없지만 비슷한 규모의 주변 건물로 인해 빈 터나 공간이 많았다. 우리는 이러한 빈 공간을 건물 안으로 들여와 작은 면적에도 불구하고 편하게 건물 안팎을 드나들 수 있는 자유로운 구조의 집을 설계해 보기로 했다.
건물은 총 4개 층으로, 지하, 1층, 2층과 옥상으로 이루어져 있다.
우리는 도심 주거 지역에서 건물을 설계할 때마다 옥상을 잘 활용하려고 노력한다. 엄격한 도시 건축 규정 때문에 건축물의 크기는 종종 제한되기 마련이다. 그 결과 비슷한 규모와 높이의 건물이 같은 한 지역에 줄지어 서게 된다. 옥상 공간을 효과적으로 잘 활용하면 도시 환경이 지닌 기회와 가능성을 극대화할 수 있다. 도심 주거 지역이라는 위치에도 불구하고 목제 테라스를 들어올린 듯한 옥상에서는 주변 경관이나 불어오는 바람을 통해 자연과 계절의 변화를 느낄 수 있다.
2층 거실에는 과장된 크기의 창문이 있다. 쌍여닫이창문을 열어 젖히면 자연 채광과 환기가 이루어지며 야외 공간을 실내로 들여온 듯한 효과까지 얻을 수 있다. 그뿐만 아니라 옥상은 옥외 계단으로 바로 연결되기 때문에 거주자는 거실에 있을 때 머리 위로 한 층이 더 있는 듯한 안전함과 안정감을 느낄 수 있다.
거주자는 지표면에서 가까운 1층이나 지하에서 항상 땅 또는 흙과 가까이 있다는 느낌을 받을 수 있다. 출입구와 흙바닥이 있는 1층은 실내와 외부 사이에 낀 듯한 분위기를 지녔으며 지하층에서는 드라이 에어리어를 통해 어느 정도 야외 공간을 느낄 수 있다. 이 공간들은 특정 목적이나 용도 없이 널찍하게 주어져 생활 방식이 변경함에 따라 유연하게 이용할 수 있다.
우리는 도심 내 좁은 부지에 소형 주택을 짓더라도 각각의 성질과 기능에 따라 각 층에 공간적 역할을 명확히 부여한다면 건물과 주변 주택가, 또는 건물의 내외부가 자연스럽게 관계를 맺고 연결될 수 있다고 믿는다.

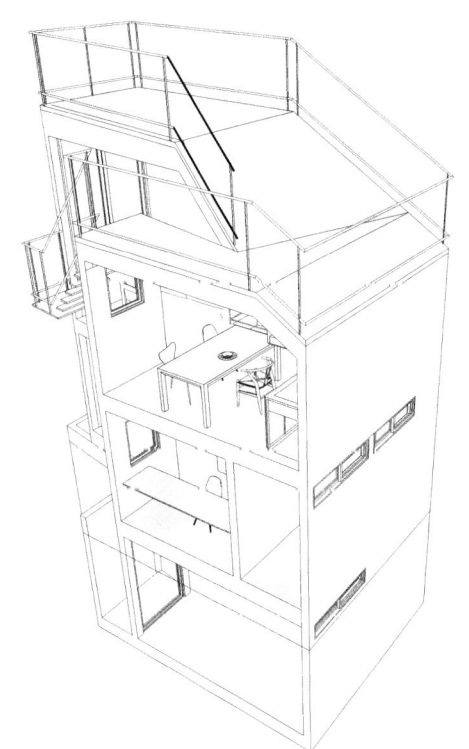

CONCEPT IMAGE

$15m^2 - 25m^2$

For maximizing the opportunities and possibilities of such urban areas, rooftop becomes a very effective space to utilize. The rooftop, as if uplifted from a wood deck, allows residents to feel the nature and transition of seasons by the surrounding scenery or the wind going through the rooftop, even though the building is situated within an urban residential area.

While the site of this building is small—about 60 sqm in size, applied with strict regulations—it is a commonly found small lot in a residential area in Tokyo.

Although it is not allowed to build a large building according to the regulation, there are many blank spaces or voids created by surrounding buildings restricted in a similar scale. By bringing in such a void into this building, we thought about producing an unconstrained house even for its minute scale, providing a sense of freedom to go in and out of the building at will.

The building is configured in 4 layers, with a basement, 1st floor, 2nd floor, and a rooftop.

We have always considered about making a good use of rooftop if designing a building in an urban residential area. Building size is often constrained by regulations in a city block applied with strict legal requirements. As a result, buildings with similar scale and height line up around such areas. For maximizing the opportunities and possibilities of such urban areas, rooftop becomes a very effective space to utilize. The rooftop, as if uplifted from a wood deck, allows residents to feel the nature and transition of seasons by the surrounding scenery or the wind going through the rooftop, even though the building is situated within an urban residential area.

There is an almost oversized opening for a living room on the 2nd floor level. When it opens by the double swinging windows, the opening brings in the daylight and wind into the room, and even creates effect as if bringing in the exterior space into the building. Furthermore, the rooftop is directly connected through an exterior staircase, so that the residents would feel a unique sense of security and stability within the living room as if there'd be another ground surface above the head.

At the layers near the ground level—on the 1st floor or the basement, the residents can always feel near to the ground or the soil. The 1st floor level with an entrance and earthen floor has a sense of in-between space of interior and exterior, and the basement level allows residents to feel outside loosely through a dry area. Those are provided as generous spaces without limiting the purposes and usages, in order to flexibly apply changes in lifestyles.

We thought a building and its surrounding residential area, or the inside and the outside of the building, can be comfortably related and connected, even through a small house built on a minimum lot within an urban area, if each layer plays a particular role with its own nature and functionality.

15m² - 25m²

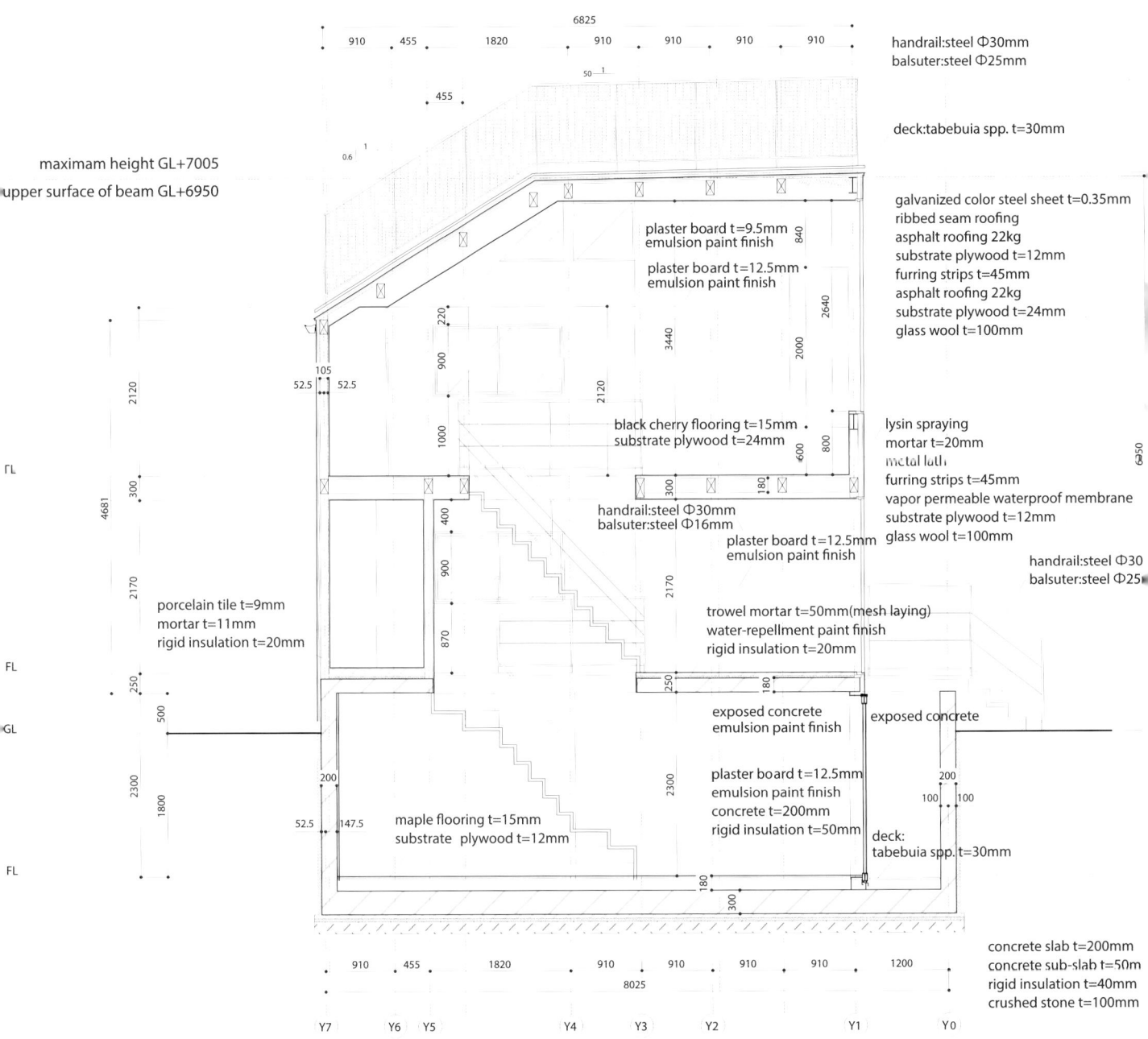

Rooftop house

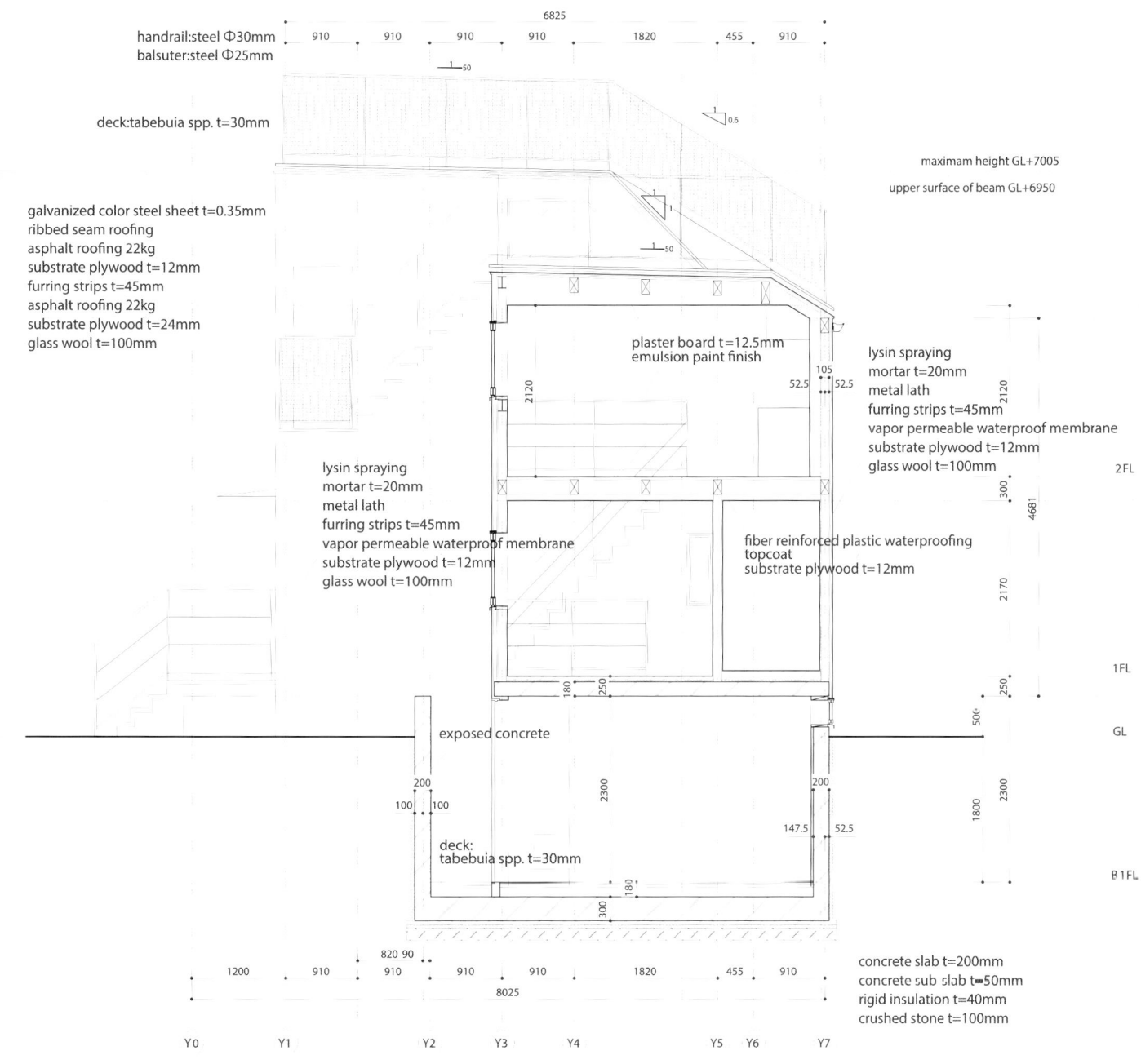

SECTION DETAIL

15m² - 25m²

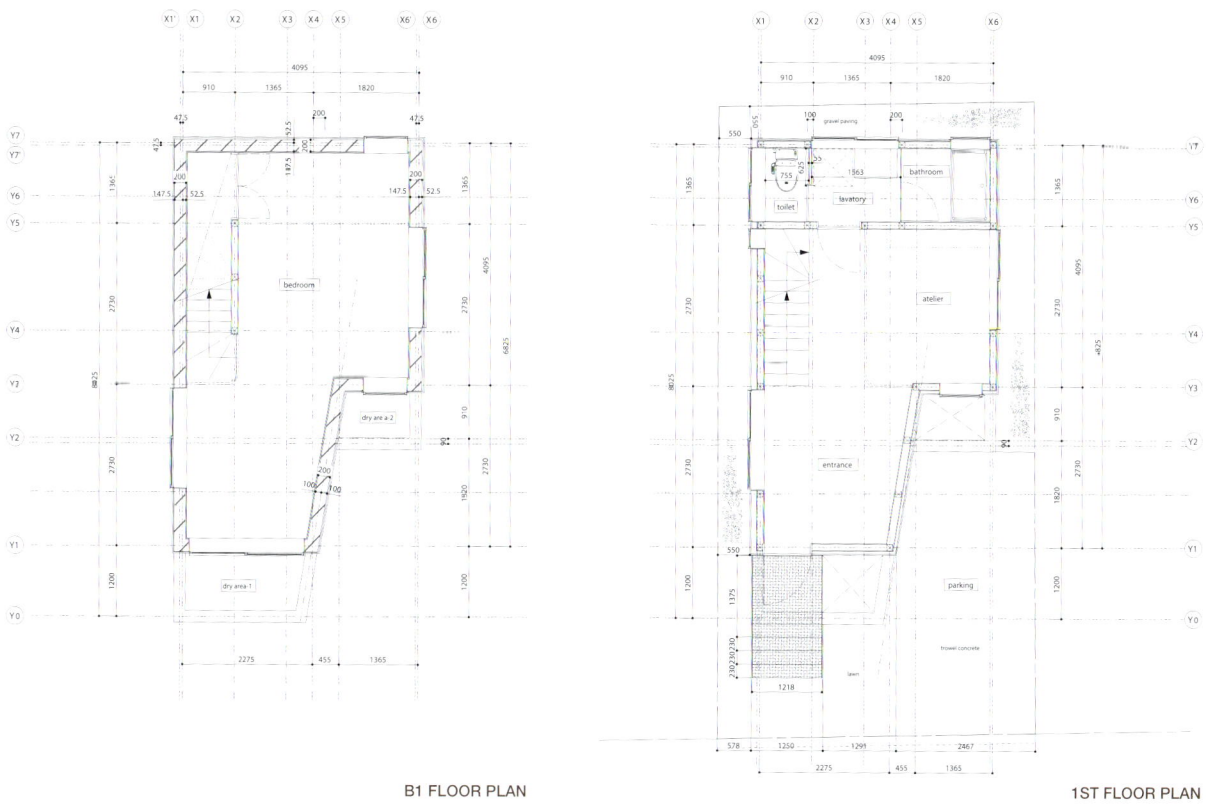

B1 FLOOR PLAN

1ST FLOOR PLAN

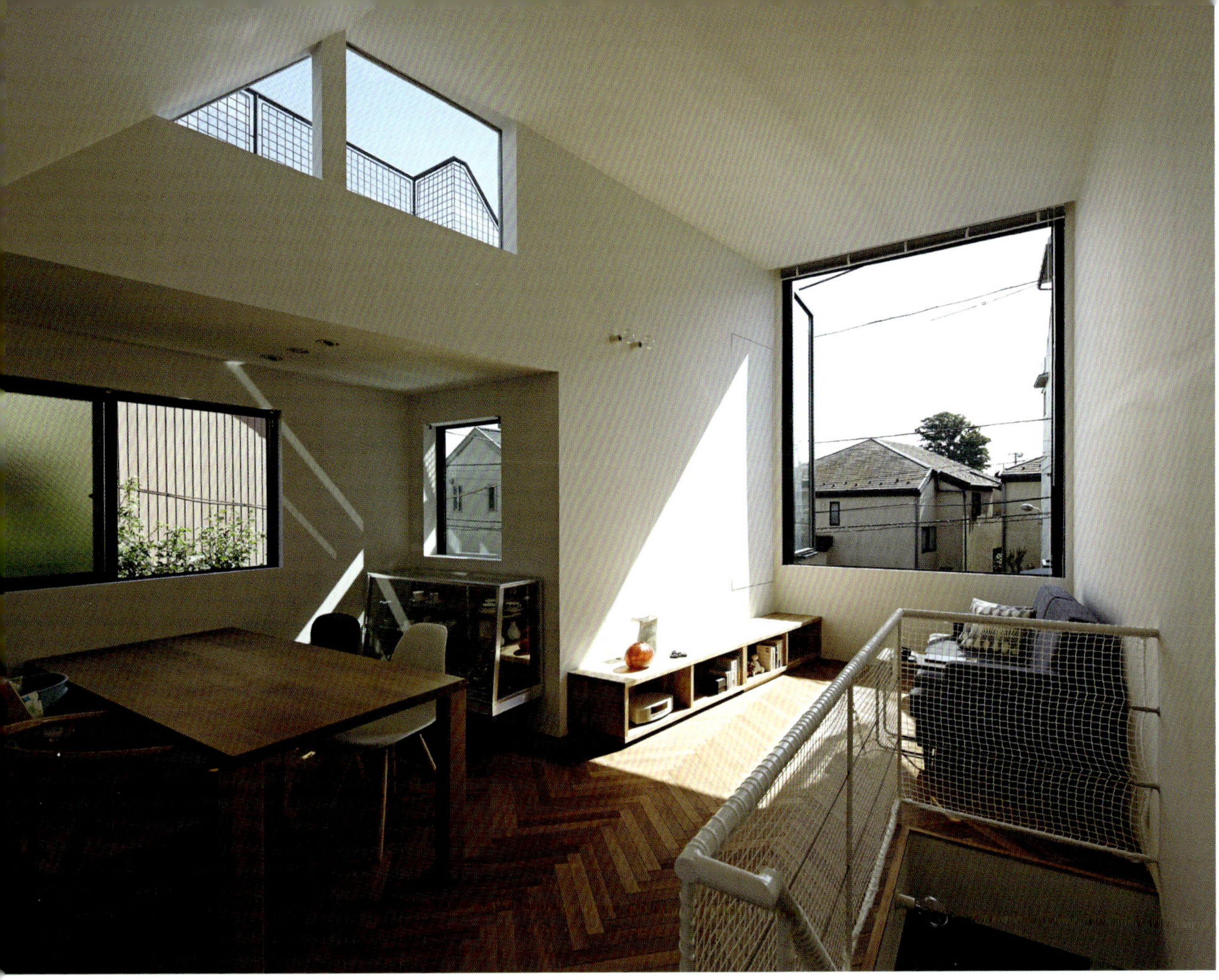

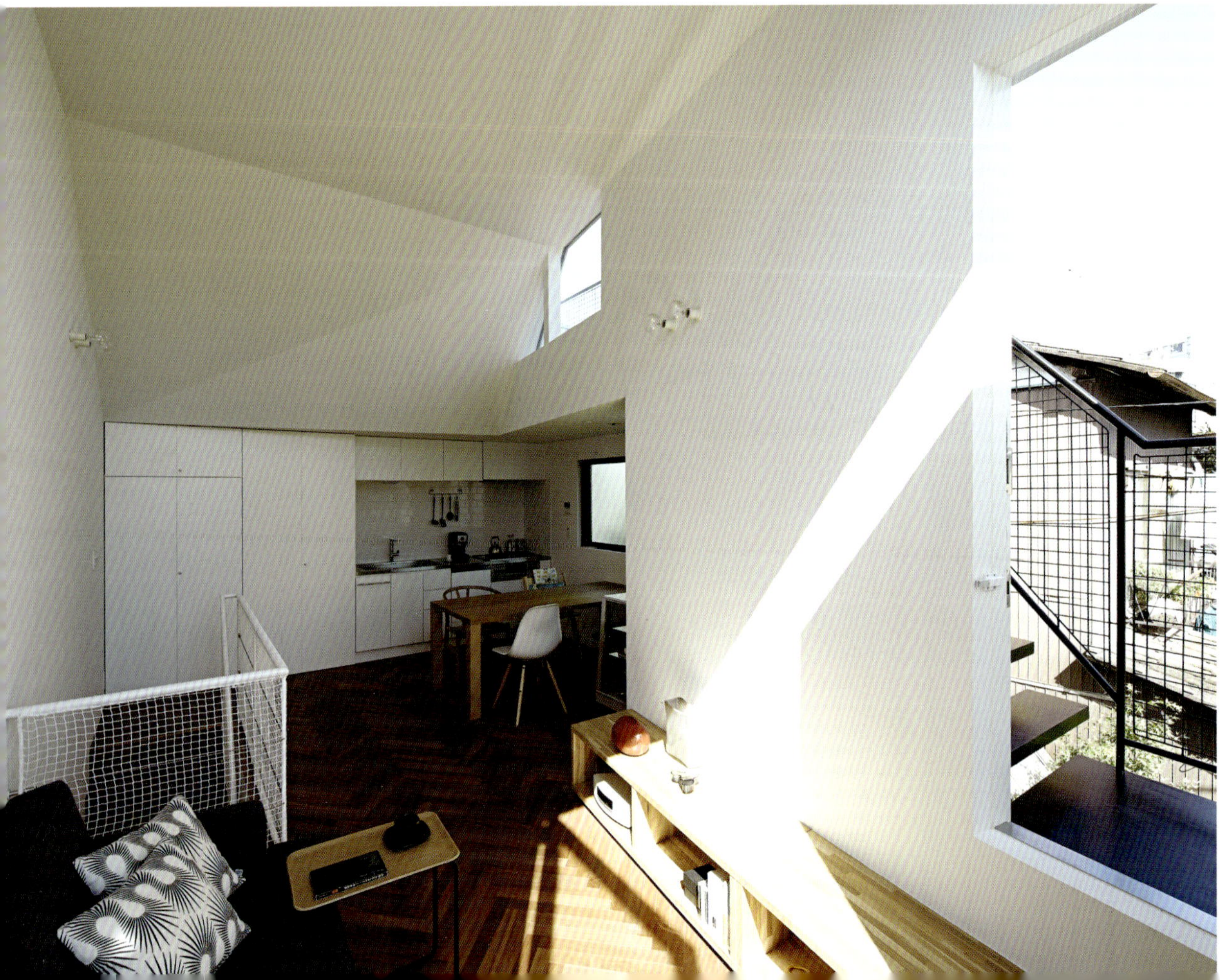

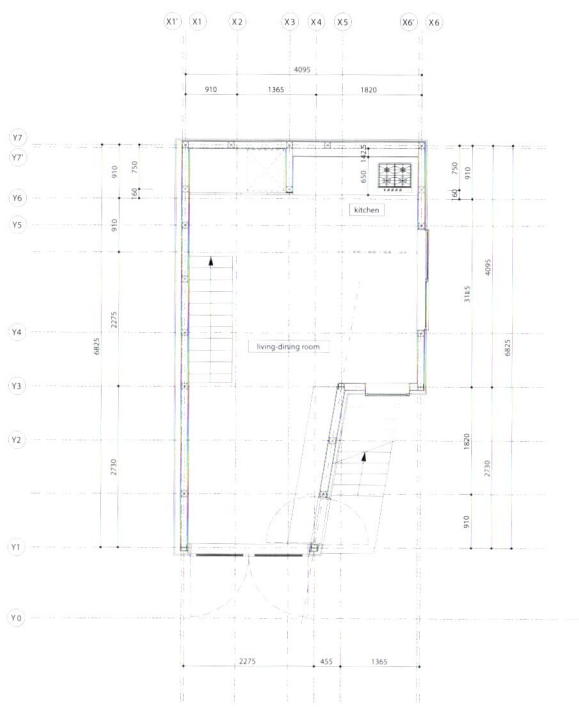
2ND FLOOR PLAN

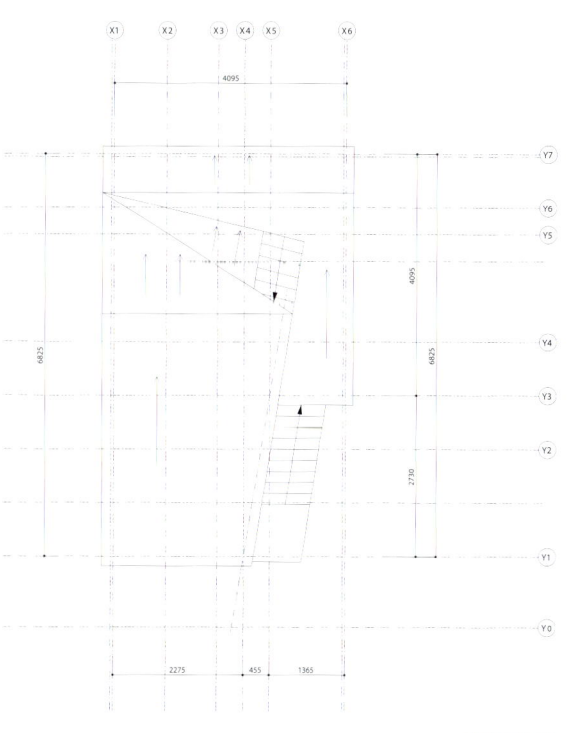
ROOF PLAN

41

$15m^2 - 25m^2$

Rooftop house

$22.87 m^2$

Long window Sunlight Spiral stair

Location
Tokyo, Japan
Use
House
Site Area
58.64m²
Built Area
22.87m²
Total Floor Area
64.42m²

Floor
B1 - 2F
Exterior Finish
Metallic siding
Interior Finish
Cloth

Project Architect
Tsuyoshi Kobayashi
Construction
Show-Yo
Photographer
Koichi Torimura

Long Window House

긴 창이 있는 집

another APARTMENT

15m² - 25m²

Long window house

58m² 남짓한 대지에서 개구부를 마련하는 일이 주요 사안이 되었다. 결과적으로 우리는 건물 남쪽과 천장에 개구부를 최대한 배치해 자연 채광, 자연 환기 및 전망을 해결했고 도로 및 이웃집과 맞닿아 있는 북쪽·동쪽·서쪽에는 개구부가 없는 구조로 설계했다.

세 명 가족을 위한 도쿄 주택가의 집.

클라이언트는 프라이버시를 보호하기 위해 닫아 두어야 하는 창문 대신 적당한 범위 내에서 열리는 창문을 설계해 자유롭고 편안한 공간을 마련해 달라고 요청했다. 그리하여 가옥으로 둘러싸인 58m² 남짓한 대지에서 개구부를 마련하는 일이 주요 사안이 되었다.

결과적으로 우리는 건물 남쪽과 천장에 개구부를 최대한 배치해 자연 채광, 자연 환기 및 전망을 해결했고 도로 및 이웃집과 맞닿아 있는 북쪽, 동쪽 서쪽에는 개구부가 없는 구조로 설계했다. 이러한 결정에 맞추어 집을 동서로 길게 뻗은 형태로 설계한 후 대지의 북측에 배치했다.

2층은 동서로 긴 하나의 방으로 구성하고 남쪽에는 벽면을 가로지르는 긴 창을 배치했다. 방은 미닫이 문으로 자유롭게 분할할 수 있으며 미닫이 문은 벽으로 다시 밀어 넣을 수 있다.

1층의 거실은 지표면보다 약간 높게 배치했고 전체가 넓은 베란다 같은 느낌을 준다. 지하에는 영상실이 있어 가족 모두가 즐길 수 있다. 공간의 중앙에는 세 개 층을 관통하는 나선형 계단이 있어 집 전체에 가벼운 일관성을 부여한다.

The selection of openings became a big theme for this 58m² -sized site surrounded by residential houses. Consequently, we adopted a plan to make maximum use of openings on the south face and the roof for natural illumination, ventilation, and views and constructed a structure without openings on the north, east, or west faces that are surrounded by a frontal road and adjacent houses.

A house for a family of three located in a residential area of Tokyo. The client requested that they needed no windows that must be kept closed in order to secure privacy but desired those that open only to a modest range to realize unconstrained comfortable space.

Therefore, the selection of openings became a big theme for this 58m²-sized site surrounded by residential houses. Consequently, we adopted a plan to make maximum use of openings on the south face and the roof for natural illumination, ventilation, and views and constructed a structure without openings on the north, east, or west faces that are surrounded by a frontal road and adjacent houses.

Along with this decision, we placed the house on the north side of the site in a shape stretching east to west.

The second floor consists of one room stretching east to west with a full of openings on the south face. The room can be divided freely with a sliding door that is retreatable into the wall.

The living room on the first floor is located a little higher than the ground level and has an atmosphere like a broad veranda as a whole. In the basement, the family can enjoy a home theater. A spiral staircase penetrating the center of these unique three layers gives a light sense of unity to the whole house.

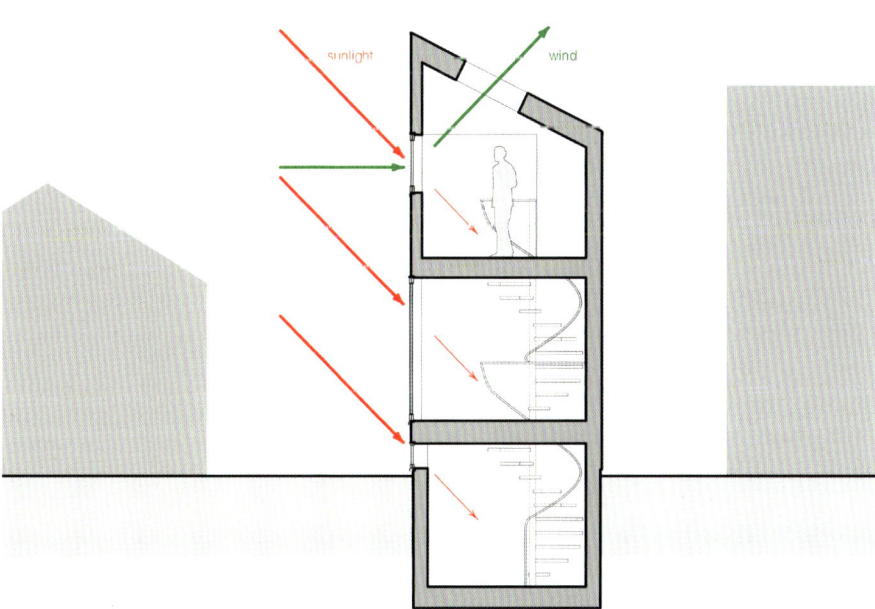

CONCEPT IMAGE

15m² - 25m²

upper surface of beam GL+ 6822

upper surface of beam GL+ 5637

2FL

1FL

GL

B1FL

Long window house

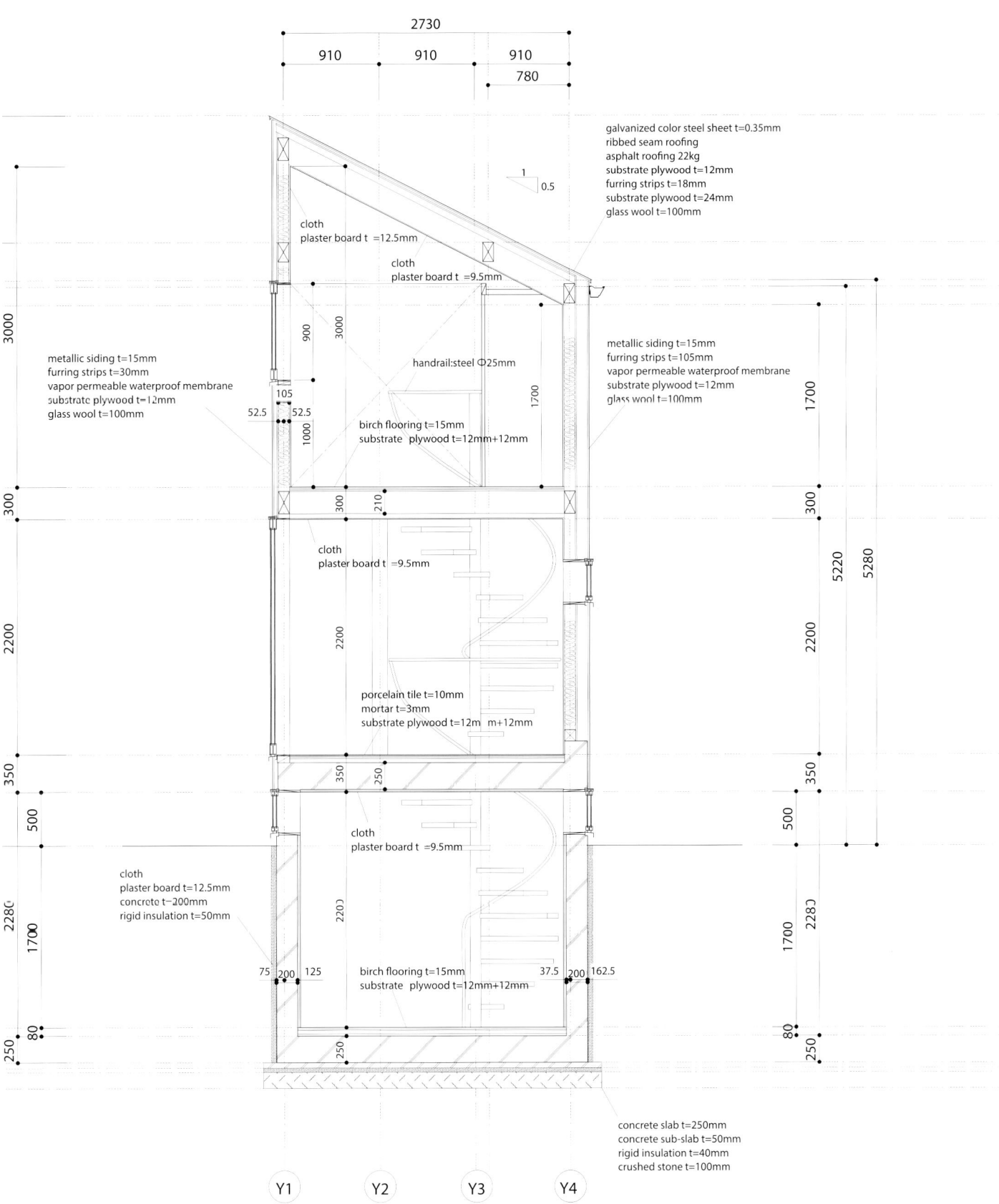

SECTION DETAIL

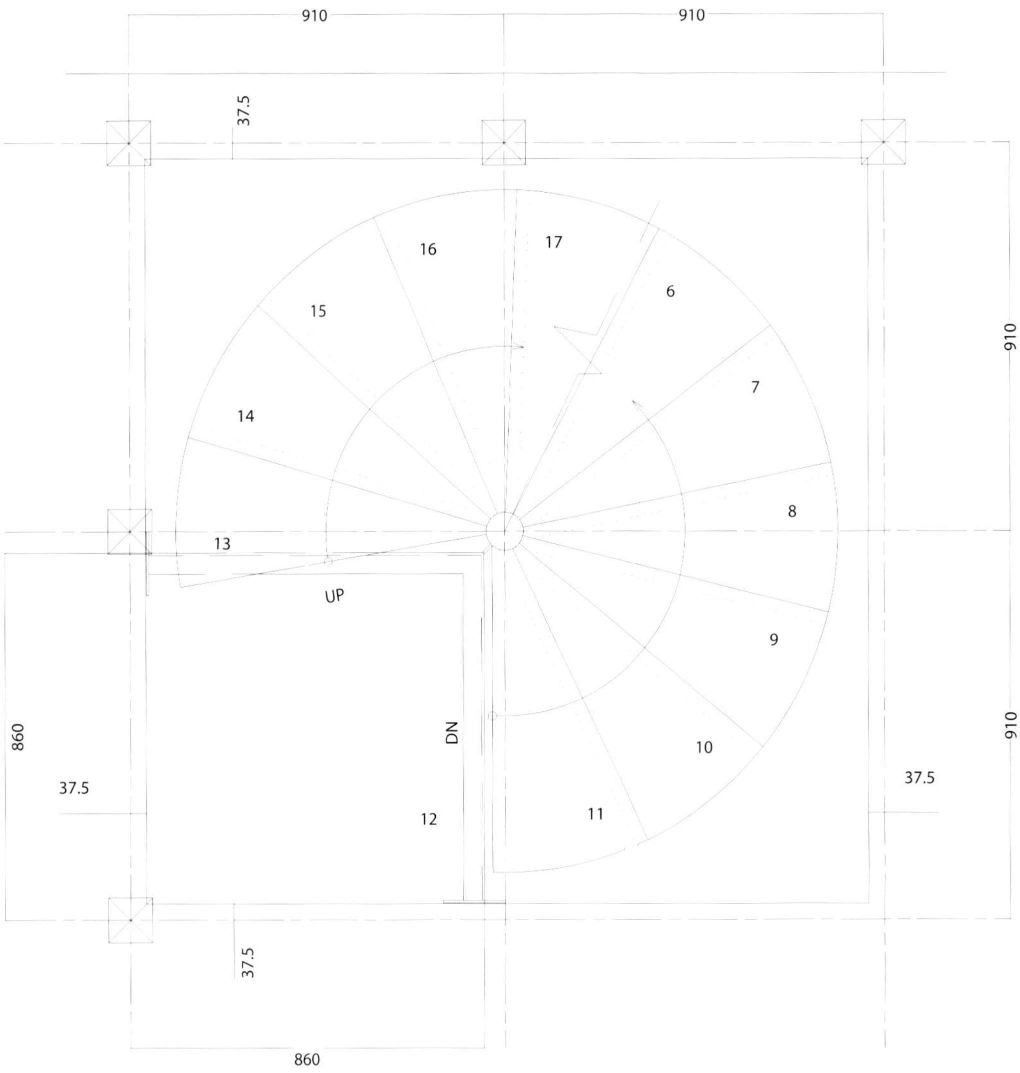

SPIRAL STAIRCASE PLAN

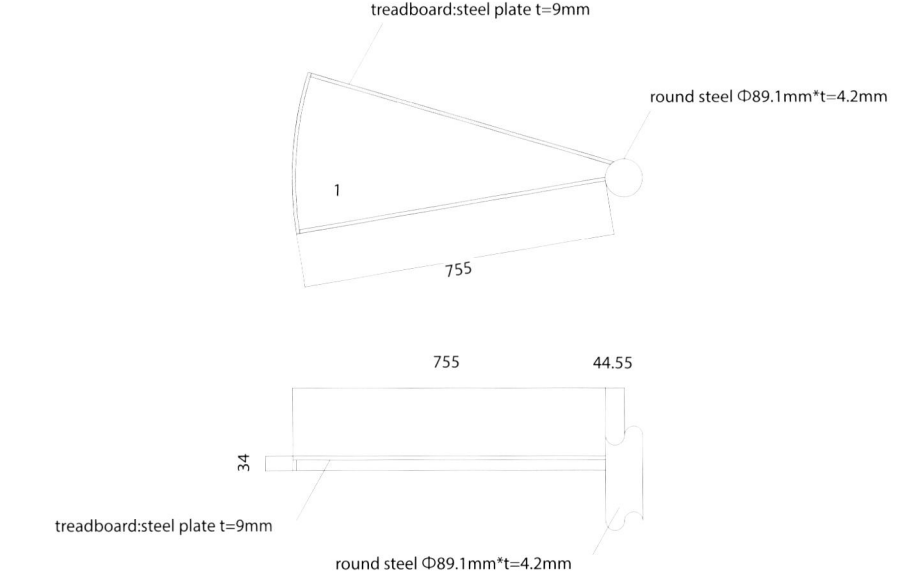

TREADBOARD DETAIL

Long window house

15m² - 25m²

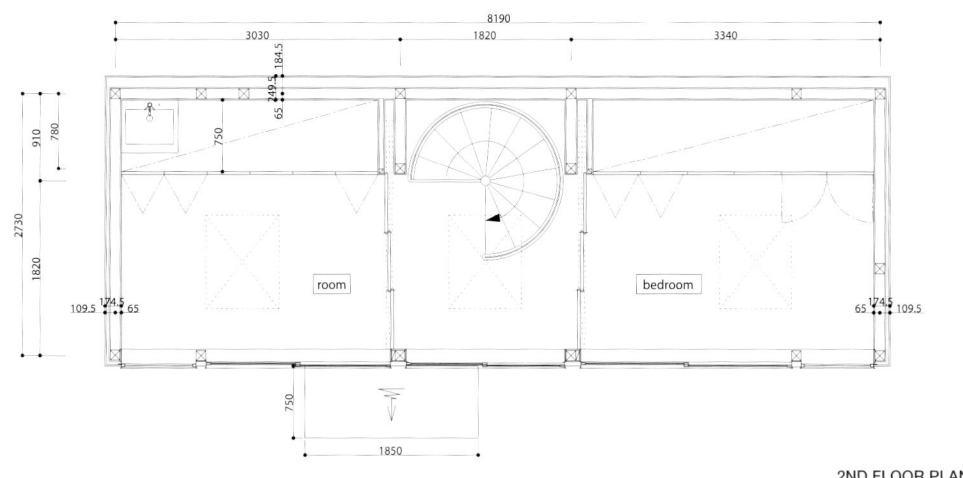

2ND FLOOR PLAN

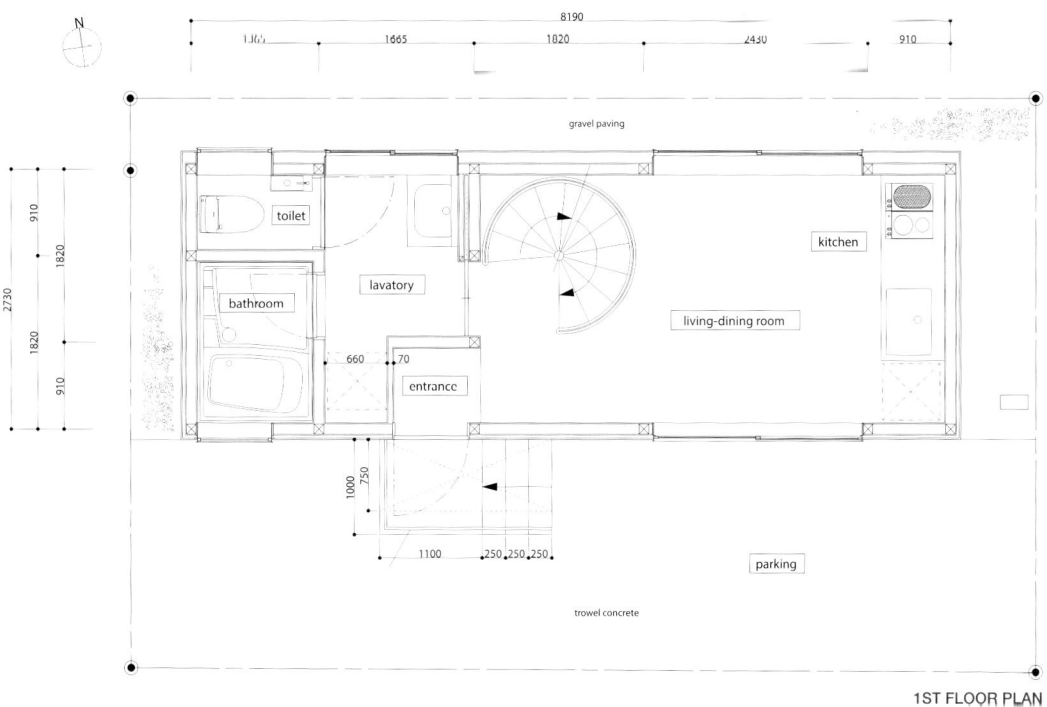

1ST FLOOR PLAN

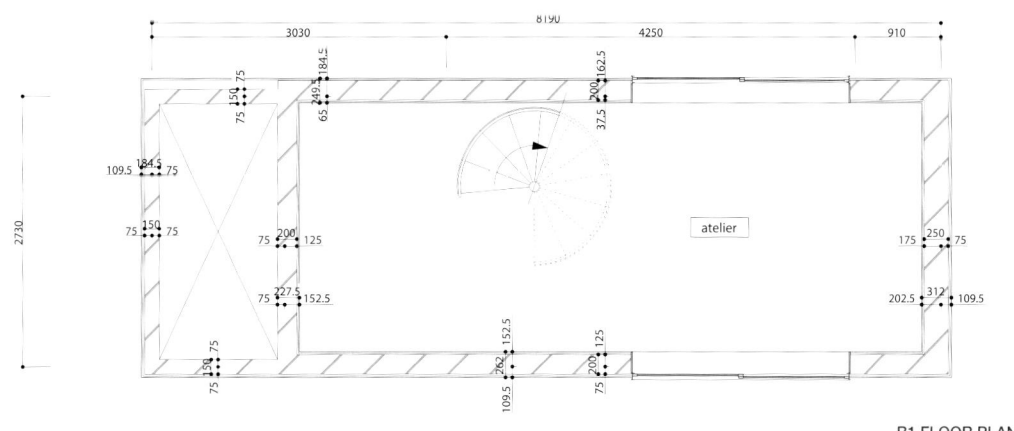

B1 FLOOR PLAN

$25.94 m^2$

Narrow
Top light window
Spiral staircase

Location
Tokyo, Japan
Use
House
Site Area
43.42m²
Built Area
25.94m²
Total Floor Area
69.29m²

Floor
3F + Roof terrace
Exterior Finish
Galvanized steel sheet, larch structural plywood
Interior Finish
Vinyl cloth, gypsum board

Project Architect
*Yukinobu Nanashima,
Tomomi Sano*
Photographer
Shinsuke Kera / URBAN ARTS

House at Hommachi

천창을 통해 환기와 채광을 극대화한 호마치 주택

atelier HAKO architects

$15m^2 - 25m^2$

주변 건물에 둘러싸인 환경에서 주거인의 생활 편의를
배려하기 위해 식사/주방 공간 위로 보이드를 덮고 있는
북측 지붕에 천창을 달았다.
그 결과 천창을 통해 들어온 자연광이 주택 북측의 깊은 공간까지
밝혀주며 공기를 위아래로 순환시켜 건물 전체를
자연적으로 환기시켜준다.

이 주택은 가로 4m 남짓한 작은 대지에 자리하며 남쪽으로는 좁은 도로와 맞닿아 있다. 이 건물은 작은 도시 블록 정도 규모의 주거밀집지역에 위치해 있는데 도로가 있는 남쪽을 제외한 나머지 3면이 이웃 건물과 비좁게 마주하고 있다.
본 프로젝트는 제한적인 면적 내에서 가족이 생활할 수 있도록 최대한 넓은 공간을 확보하는 것이 목표였다. 전체적인 구조는 좁은 공간을 최대한 활용했으며 용도에 따라 천장의 높이를 달리한 세 개 층으로 구성되었다.

식사 공간 위로 보이드를 배치하고 1층에 다용도 옥외 현관을 마련해 건물 용적 내에서 내부 공간을 확장할 수 있도록 설계했다.

다음으로 주변 건물에 둘러싸인 환경에서 수거인의 생활 편의를
배려하기 위해 식사/주방 공간 위로 보이드를 덮고 있는 북측 지붕에
천창을 달았다. 그 결과 천창을 통해 들어온 자연광이 주택 북측의 깊은
공간까지 밝혀주며 공기를 위아래로 순환시켜 건물 전체를 자연적으로
환기시켜준다
또한 여러 방향으로 트인 공간을 설계하고 각 층을 나선형 계단에 연결된
요소로 배치함으로써 실내 공간 내에서 동선의 효율성을 높였다.
이 주택의 실내 공간에서는 수직적, 수평적 요소가 역동적인 조화를 이루어
재미있고 독특한 연속성과 장애물 없이 트인 공간감을 느낄 수 있다.

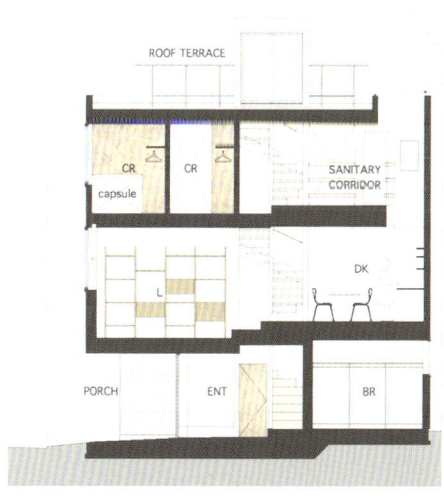

SECTION

$15m^2 - 25m^2$

House at Hommachi

In order to be able to get a level of amenity for dwellers, we set the top light windows on the north side of the roof above the void over the dining kitchen space.
As the result, the top light windows offer the sufficient natural light to the depth in the north side of the house and offer the natural ventilation whole of the building utilizing the temperature difference between the upper and lower air.

This house sits on a tiny site that has the only approximately 4 meters width, facing the narrow road on the south side. In this crowded dwelling area of the small city block, this house faces the neighbor buildings built close to the site boundary around this house on the three directions except the south road side.
It was inserted in a target to reserve the maximum useful living space for the family in the limited site. An overall frame was provided that include the three layers of space that have the area of the building coverage very limit and each different ceiling height according to the require.
The surplus area over building volume limit was solved, by being replaced to the void over the dining space and being replaced to the outside multi-purpose porch on the ground floor, as factors to bring an expanse in the interior space.
Then, in this surrounded environment, in order to be able to get a level of amenity for dwellers, we set the top light windows on the north side of the roof above the void over the dining kitchen space. As the result, the top light windows offer the sufficient natural light to the depth in the north side of the house and offer the natural ventilation whole of the building utilizing the temperature difference between the upper and lower air.
And we tried to create fluidity of the internal space by providing the openings toward the various directions, and providing the floors as continuous elements to the spiral staircase.
The dynamism accompanied by the mixture of horizontal direction and vertical direction in the interior space of this house resulted to unique enjoyable sequence and spacious barrier-free feeling.

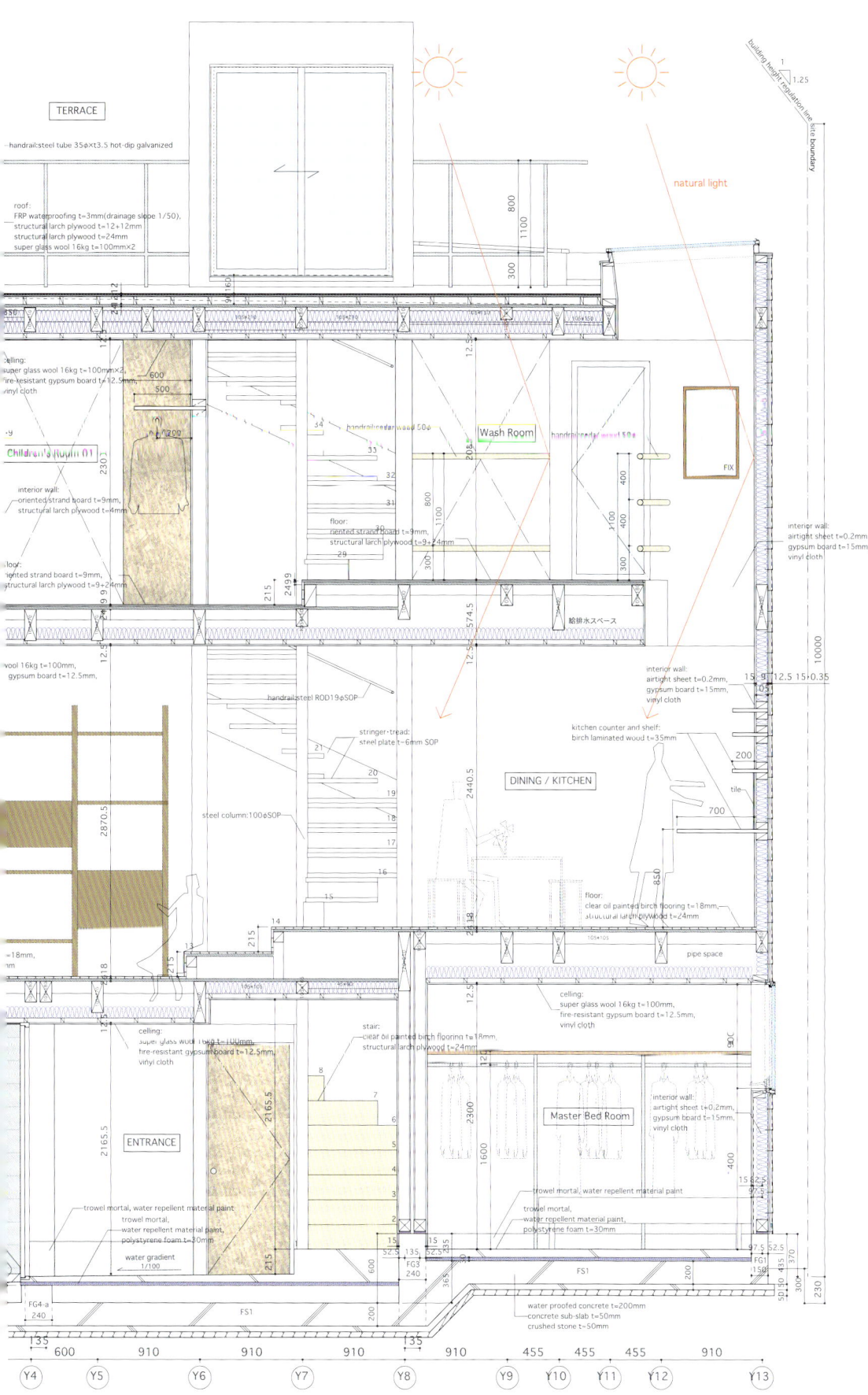

SECTION DETAIL

$15m^2 - 25m^2$

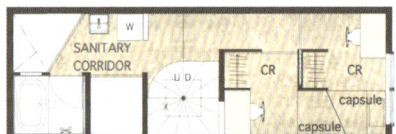

UPPER FLOOR PLAN

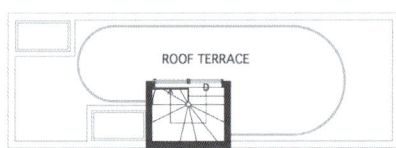

ROOF FLOOR PLAN

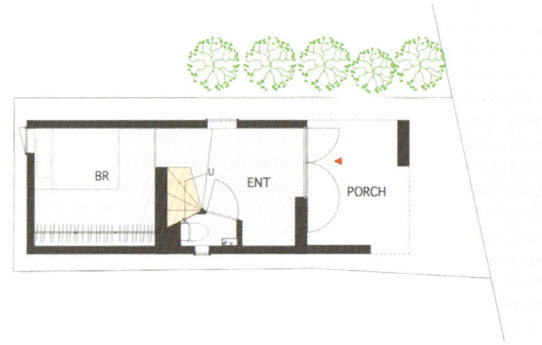

GROUND FLOOR PLAN

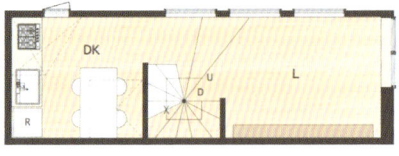

MAIN FLOOR PLAN

atelier HAKO architects

interview

자연채광은 협소한 공간에서 건강한 생활을 위한 솔루션으로, 이를 풍부하게 활용할 수 있는 여러 방안을 연구했다. 또한, 좁고 긴 형태에서 자연 환기를 극대화할 수 있는 건축적 요소로 천창을 사용하였다.

Q. 클라이언트의 요구사항은 무엇이었으며, 그에 따른 건축적 대응은 어떤 것 이었나?
A. 클라이언트는 극도로 좁은 면적 내에서 건강하고 품위 있게 생활할 수 있는 주거 공간을 의뢰해왔다. 그 요청 내용에 부응하기 위해 우리는 자연 채광을 풍부하게 활용할 수 있는 여러 방안을 연구했다.

Q. 협소한 대지 위, 수납공간을 창출하기 위한 아이디어는 어떤 것이었으며, 수직적으로 프로그램 공간구성을 이떻게 하였는가?
A. 이용 편의성을 확보할 수 있는 적절한 크기를 파악하고, 수직 방향으로 시각적 거리감과 실제 공간의 연속성 사이에서 균형을 이루는 부분에 중점을 두었다.

Q. 예산을 줄이거나, 제한된 예산 안에서 최고의 퀄리티를 뽑아내기 위한 아이디어는 무엇인가?
A. 비용이 전혀 들지 않는 자연광, 그림자, 자연풍 및 시간에 따라 달라지는 이들 요소의 변화를 최대한 잘 활용하는 것이 중요하다고 생각한다.

Q. **What were your client's requirements and your architectural approach for them?**
A. The client asked us for a living space where they can live healthily with dignity in the ultimately small site. In response to the request, we explored ways to lead rich natural light into the room.

Q. **What were your ideas to utilize the space & storage on small site and how to zone the program in a vertical way?**
A. We believe that it is important to explore suitable sizes according to ease of use and to arrange the balance between visual distances and actual continuity in the vertical direction.

Q. **Within a limited construction budget, would you be able to share your tips to make a higher quality project?**
A. We think that it is the most important to make the best use of the costless elements such as natural light, shadow, natural wind, and movements of them as time goes onnt on one house and at the same time raise the quality and thought put into the design.

23m²

Toplight Various height Rooftop terrace

Location
Buenos Aires, Argentina
Use
House & office
Built Area
23m²
Total Floor Area
89m²

Exterior Finish
Curved sheet metal prepainted white
Interior Finish
Sheetrock

Project Architect
Constanza Chiozza, Pedro Magnasco
Construction
Patricio Construye & Oscar Ojeda Construcción
Photographer
Javier Agustín Rojas

PH Lavalleja

기하학적인 구조로 외관의 연속성을 살린 PH 라바예하

CCPM Arquitectos

주변 건물에서 이 주택이 솟아 있는 형상을 바라보면 블록의 내부에 공공 공간을 마련할 수 있다는 가능성도 엿볼 수 있다. 주택 개조 과정에서 기존의 지붕을 해체하고 대신 하나의 외장재가 건물 외관을 둘러싸며 서로 다른 공용 공간을 연결하고 건물의 내외부가 조화를 이루도록 시공했다.

PH는 부에노스 아이레스 전통 가옥의 한 유형으로 고밀도 저층구조가 특징이다. 길다란 주거지역의 끝자락에 자리한 PH 라바예하(PH Lavalleja)는 주변의 주거용 고층 프리 플랜(free plan) 건물에 둘러싸여 있다. 주변 건물에서 이 주택이 솟아 있는 형상을 바라보면 블록의 내부에 공공 공간을 마련할 수 있다는 가능성도 엿볼 수 있다. 주택 개조 과정에서 기존의 지붕을 해체하고 대신 하나의 외장재가 건물 외관을 둘러싸며 서로 다른 공용 공간을 연결하고 건물의 내외부가 조화를 이루도록 시공했다. 기존 건물의 구조와 환경적 제약은 기하학적인 구조로 풀어내어 외관의 연속성을 살렸다. 실내에는 목재 마감과 가구를 반복적으로 배치해 재질감의 변화로 공간을 구분했다.

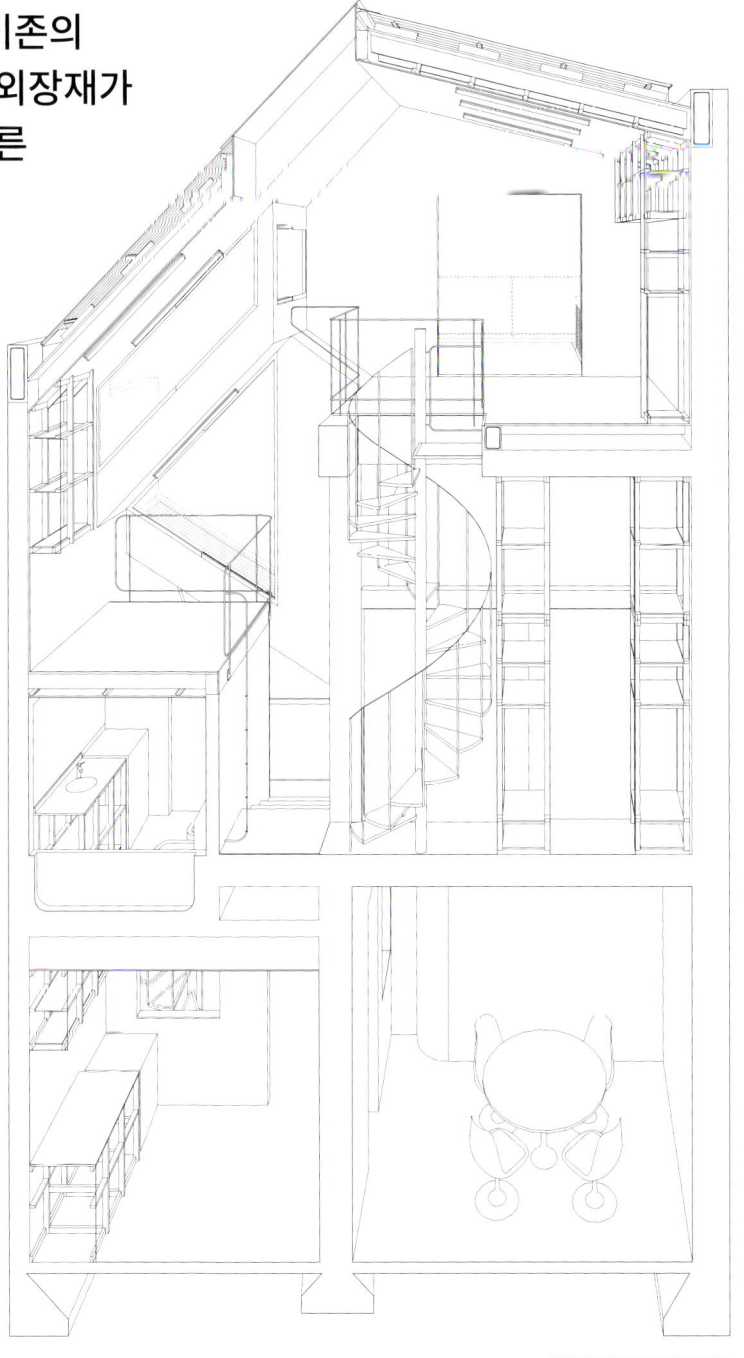

PERSPECTIVE INTERIOR

15m² - 25m²

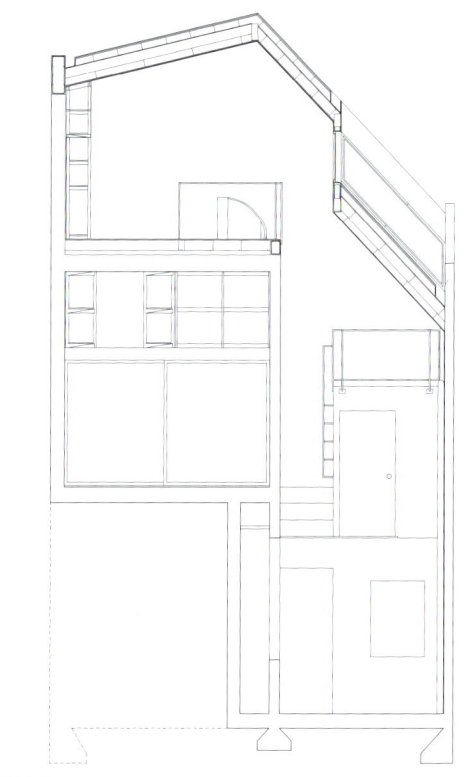

SECTION A

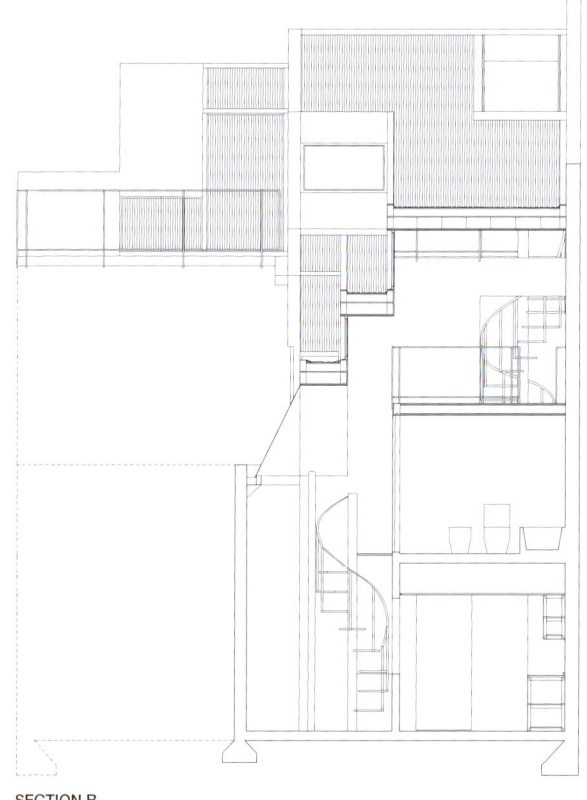

SECTION B

The views from them frame the scenario in which architecture emerges, opening possibilities for public space in the interior of the block. The update consists in disassembling the existing roofs and replacing them with a continuous envelope that links the different public spaces and mediates between interior and exterior.

PH is the name given to a traditional housing typology in Buenos Aires, characterised by its high density and low rise. Set in the last unit of a long plot, PH Lavalleja coexists with the neighbouring free plan, high rise residential buildings that surround it. The views from them frame the scenario in which architecture emerges, opening possibilities for public space in the interior of the block. The update with a continuous envelope that links the different public spaces and mediates between interior and exterior. The rhythm of the existing infrastructure and the perimetral constraints are worked geometrically to generate the continuity of the envelope. A system of wooden surfaces and furniture is replicated throughout the interior, generating different situations by shifting materialities.

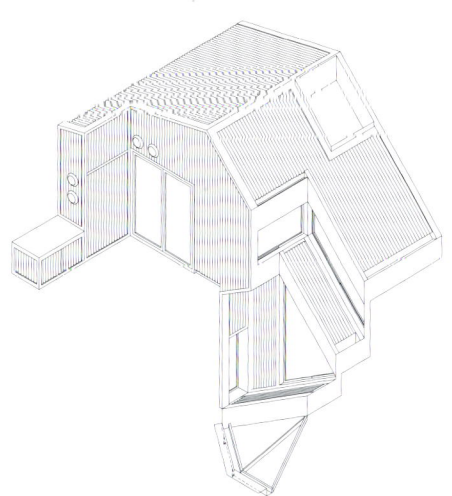

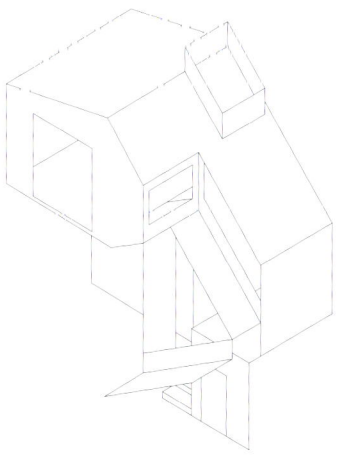

AXONOMETRIC. EXTERIOR ENVELOPE - CORRUGATED METAL

AXONOMETRIC. INTERIOR ENVELOPE OSB/SHEETROCK

15m² - 25m²

PH Lavalleja

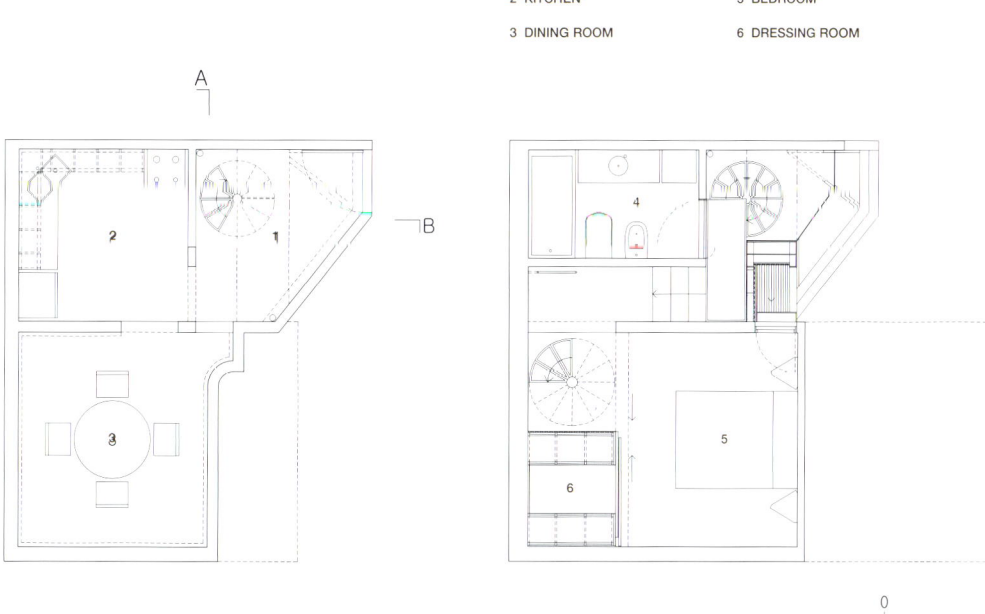

1 HALL
2 KITCHEN
3 DINING ROOM
4 BATHROOM
5 BEDROOM
6 DRESSING ROOM

GROUND FLOOR PLAN

1ST FLOOR PLAN

1. CURVED SHEET METAL PREPAINTED WHITE
2. ZINKED CUSTOM FOLDED SHEET METAL WITH OIL BASED WHITE PAINT
3. "O" TYPE LIGHT STEEL FRAME PROFILE - 15MM
4. "C" TYPE LIGHT STEEL FRAME PROFILE - VARYING SECTION
5. THERMAL INSULATION (EPS) 30MM
6. DUPONT TYVEK TYPE MEMBRANE
7. ORIENTED STRAND BOARD (OSB) 12MM
8. PROJECTED CELLULOSE INSULATION
9. COMPOSED LIGHT STEEL FRAME TUBE (C + U TYPE PROFILES) - VARYING SECTIONS
10. WINDOW COMPOSED OF NORMALIZED "L" TYPE METAL WITH OIL BASED WHITE PAINT
11. SHEETROCK 12,5MM
12. 4+4MM GLAZING
13. MASONRY WALL
14. WHITEWASH
15. FINISH WHITE PAINT
16. "Z" TYPE PROFILE FOR SHEETROCK FINISHING
17. TRIANGULAR BRACKET WITH "T" TYPE PROFILE
18. WELDED POINT
19. ADHESIVE ELASTIC JOINT
20. CONCRETE BEAM (DIMENSIONS AND DOSAGE ACCORDING TO CALCULATION)
21. WHITE PVC MEMBRANE
22. POLIESTER WOVEN GEOTEXTILE
23. DRAINING LAYER(POME TYPE ROCKS)
24. SOIL
25. GRASS "GRAMA BAHIANA"
26. CONCRETE SLAB

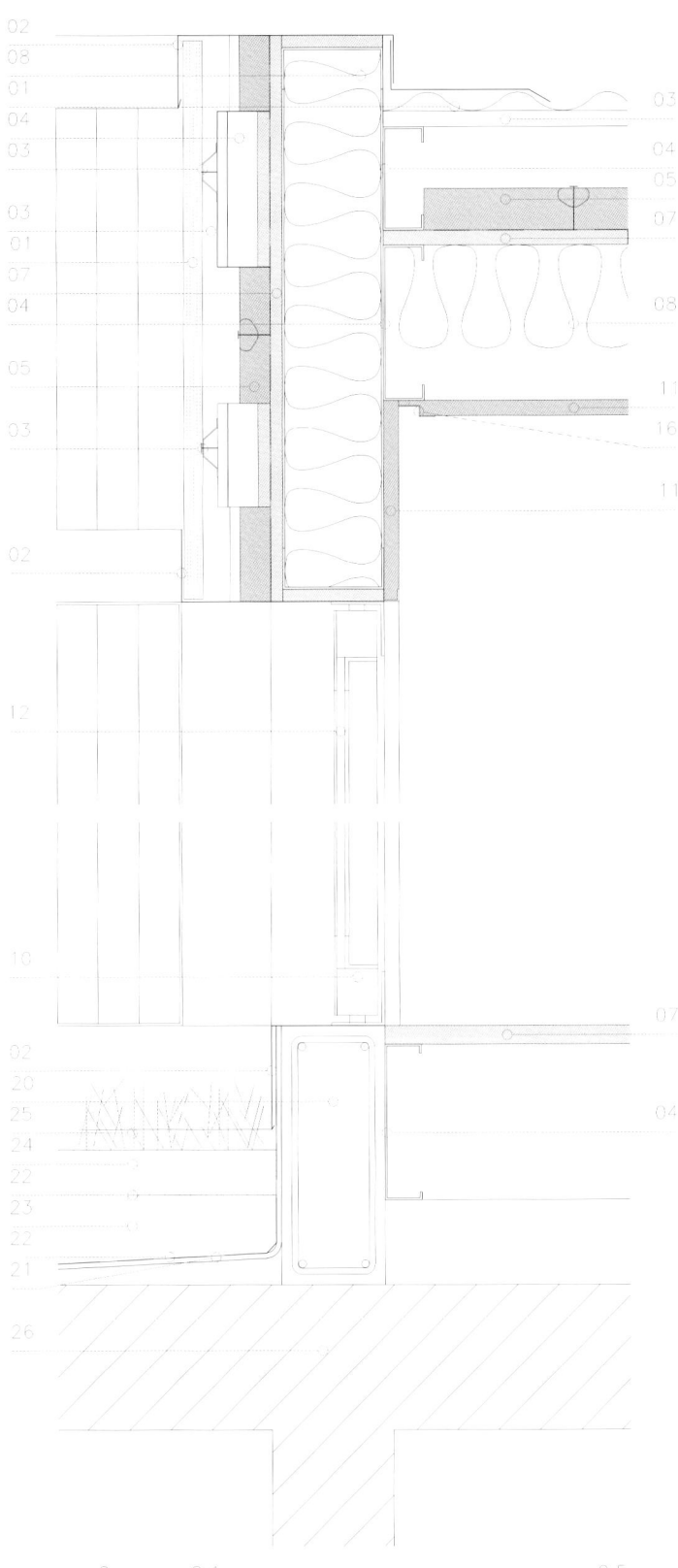

PARTIAL SECTION DETAIL

PH Lavalleja

5 BEDROOM
6 DRESSING ROOM
7 DESKTOP
8 LIVING ROOM
9 FACILITIES CABINET
10 ROOFTOP TERRACE

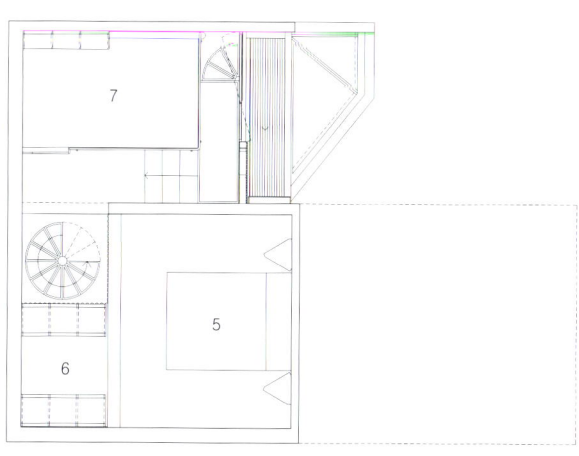
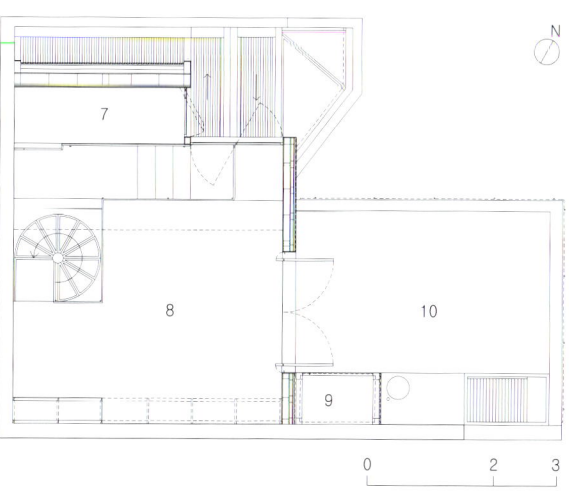

1ST FLOOR PLAN_ MEZZANINE

2ND FLOOR PLAN

CCPM Arquitectos

interview

자연광이 좋고 자연 환기가 되는, 서로 연결된 공간이자 녹지가 있는 환경을 조성하는 것이 목적이었다. 새로운 지붕 구조물을 만들고 일부 벽을 철거하여 밝고 개방된 공간을 조성했다.

Q. 클라이언트의 요구사항은 무엇이었으며, 그에 따른 건축적 대응은 어떤 것이었나?

A. 우리가 우리의 클라이언트이었다(Constanza Chiozza와 Pedro Magnasco). 우리는 부에노스아이레스의 건축대학(FADU-UBA)을 졸업한 젊은 건축가 부부로, PH를 개조하면서 전문가적인 협력 관계를 구축했고 CCPM Arquitectos를 설립했다. 이러한 조건에서 해당 프로젝트는 우리의 건축적 관심사이자 우리가 부에노스아이레스를 이해하는 데 영향을 미치는 주제였다. 건축 계획에서 우리가 관심을 둔 주제는 주거와 업무가 모두 가능한 장소를 만드는 것이었다. 프로젝트의 목적은 장소의 기능을 강조하는 데 있지 않았다. 자연광이 좋고 자연 환기가 되는, 서로 연결된 공간이자 녹지가 있는 환경을 조성하는 것이 목적이었다. PH는 시간에 따라 집이나 사무실 이상으로 사용되며 다양한 용도를 포용할 수 있을 것이다. 실제로 우리가 일하는 곳은 메자닌이지만, 사무실 확장이 필요하면 PH의 다른 곳을 사무실 업무로 활용할 예정이다.

Q. 협소한 대지 위, 수납공간을 창출하기 위한 아이디어는 어떤 것이었으며, 수직적으로 프로그램 공간구성을 어떻게 하였는가?

A. 건물 연면적이 좁은 것이 꽤 어려운 문제였다. 서로 다른 공간에 수직으로 연속성을 주는 방식을 찾아서 이 문제를 해결했다. 이번 프로젝트에서는 새로운 지붕 구조물을 만들고 일부 벽을 철거하여 밝고 개방된 공간을 조성했다. 주택을 둘러싸는 새로운 외부 구조물로 내부의 연속성을 유지했다. 프로젝트를 시작할 때부터 이러한 연속성이 기존의 폐쇄적이고 밀실 공포증을 일으킬 것 같은 공간에 활기를 불어넣고 현대적인 환경을 만들어준다고 생각했다. 우리는 이 프로젝트가 역동적으로 시간에 따라 변화한다고 생각했고, 부분과 전체 사이의 긴장감을 고려하여 창고를 전략적으로 벽에 고정시켜 만들었다. 그래서 변화를 수용할 수 있는 체계적인 접근이 가능했다.

Q. 예산을 줄이거나 제한된 예산으로 최대한의 퀄리티를 만들어내기 위한 방법은 무엇이었나?

A. 비용을 절약할 수 있는 항목을 파악하기 위해서는 콘셉트면에서 완전한 프로젝트 제안서를 미리 구상하는 일이 매우 중요했다. 이렇게 복잡한 프로젝트를 실현하는 방식에 대한 아이디어를 아주 간단하게 표현하는 일이 중요하다. 핵심 요소는 경량 철골 기술을 이용하고, 프로젝트와 그 경제적 제약에 맞춰 작업하는 방법을 찾기 위해 간결한 노하우를 얻는 것이었다. 프로젝트를 실현하는 모든 과정에서 다양한 자료가 필요하며, 디테일을 생각하는 방식 또한 체계적이어야 한다. 그렇기 때문에 구조를 공고히 하는 OSB 같은 핵심 요소를 인테리어의 마무리 요소로도 활용할 수 있었다. 이러한 아이디어를 생각하는 능력은 제한된 예산으로 수준 높은 프로젝트를 진행할 때 매우 중요하다.

Q. What are your tips to make a better quality project within a limited budget in construction method?
A. We were our own clients (Constanza Chiozza and Pedro Magnasco). We are a young couple of architects who graduated from architecture school here in Buenos Aires (FADU- UBA). The reform of this PH is what triggered our professional partnership and the creation of CCPM Arquitectos. This condition allowed the project to be a thesis of our architectural interests and their interaction with our interpretation of the city of Buenos Aires.

The programmatic interests were to create a place where we could live and work. The project never intended to highlight the function of the places, but to create a certain kind of environment with natural light, connected spaces, natural ventilation and green areas. The PH will be able to accommodate fluctuations of use in time, becoming more of a house or more of an office. The mezzanine is actually where we are working now, but if the office grows another place of the PH should accommodate the office program.

Q. What were your ideas to utilize the space & storage on such a small site and how to zone the program in a small area?
A. The small floor area of the building was quite a challenge. We addressed this by looking for the way to create vertical continuity through the different spaces. The relationship between the new roof structure and the demolition of some walls allowed the project to create this open and light spaces. Since the new envelope rests in the perimeter of the house, this allowed the continuity of the interior. From the beginning of the project we felt this continuity would make the originally very closed and claustrophobic spaces come to life and create a contemporary environment. We think the program in a dynamic manner, able to fluctuate over time. The storage was set into some strategic walls thinking of it as a tension between part and hole. Hence there is a systematic approach able to accommodate variation.

Q. Within a limited construction budget, would you be able to share your tips to make a higher quality project?
A. Conceiving a full conceptual project proposal in advance was really important to know in which items we could economize. It is very important to have a very straight forward idea of how the materialization of a project of this complexity is going to work. A key element was to have a very concise know how of the light steel framing technology and knowing how it could work favouretly for the project and its economic constraints. As every aspect in the materialization of the project, a lot of materials are needed and it is very important to be systematic in the way the details are thought. That is why a key element such as OSB that serves the purpose of the rigidization of the structure, can also be used as a finishing element of the interior. This kind of intelligence is very important for having a good quality project with a limited budget.

15m²

Terrace Sculptural

Location
Haugesund, Norway
Use
House
Built Area
15m²

Photographer
Bent René Synnevåg

Slice

정원의 한 조각 같은 예술적인 집

Saunders Architecture

Slice

검은색 외관과 흰색 인테리어의 색상 대비는 이 건물이 정원의 '한 조각(a slice)'처럼 보이게 한다. 이 주택의 핵심은 오후 시간을 보내기에도, 하룻밤 자기에도, 아침을 맞기에도 좋은 공간이어야 한다는 점이었다. 테라스를 넓게 뽑아내자 더 큰 방이 만들어졌고 조형적인 형태가 집 전체의 틀을 이루면서 마당까지 만들어냈다.

전세계적인 경기 침체와 중앙집중화의 영향으로 세계 각지의 건축가들은 소규모 주택 설계에 도전하고 있다. 이들이 그려내는 아이디어는 공간의 규모와 반비례하는 경우가 많다.

건축가 입장에서는 15m² 짜리 작은 건물이 5,000m² 면적의 거대한 건물보다는 훨씬 즐거운 작업이다. 규모와 예산이 대규모로 축소된다고 해서 딱히 창의력까지 제한되는 것은 아니기 때문이다.

고객이 요청한 독채 야외 게스트룸을 설계하기 위해 건축가와 동료들이 처음 둘러앉아 고민한 생각은 '어떻게 하면 이 프로젝트에서 최대한 효과적인 결과물을 낼 수 있을까?'였다. 이들은 주어진 대지의 전제 사항을 고려해야 했는데, 바로 정원에 심어진 오래된 자두나무를 보존하는 것이었다. 그 결과 현재 지도니무들은 삼각형의 건축물을 뚫고 나와 방해받지 않은 채 자라고 있다. 검은색 외관과 흰색 인테리어의 색상 대비는 이 건물이 정원의 '한 조각(a slice)'처럼 보이게 한다.

예술가는 이따금씩 건축가의 영역에 속하는 작업을 하고 건축가 역시 예술과 건축의 교차점에서 활동하기도 한다. 건축가는 소규모 건축에서 상상의 나래를 펼치기가 더 쉽다고 생각한다. 최소한 성공 가능성은 더 높다. 더그 에이트킨(Doug Aitken)이나 올라퍼 엘리아슨(Olafur Eliasson)과 같은 예술가 역시 소형 주택을 제작한 적이 있다. 어쩌면 조각과도 같은 형태의 소형 주택이야말로 예술과 건축의 경계에 놓여있는 것이 아닐까?

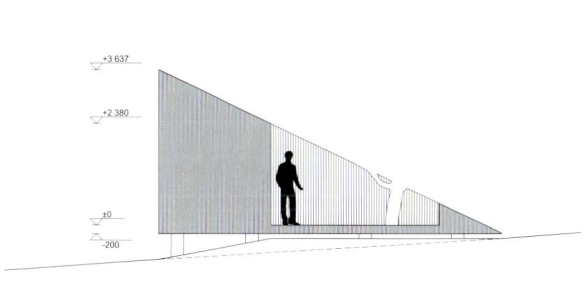

ELEVATION

'슬라이스(Slice)'는 토트 손더스가 노르웨이 헤우게순(Haugesund) 외곽의 슬라테빅(Slattevik)에 위치한 작은 정원에 설계한 소형 주택이다. 이 프로젝트의 난제는 '어떻게 하면 가장 재미있는 방법으로 15m² 짜리 주택을 설계할 수 있을까?'였다. 건물은 작지만 완벽한 디자인이어야 했다. 이 주택의 핵심은 오후 시간을 보내기에도, 하룻밤 자기에도, 아침을 맞기에도 좋은 공간이어야 한다는 점이었다. 결과적으로 '슬라이스'는 우리가 흔히 주택에 대해 갖는 기대치를 뛰어넘었고 훨씬 독창적이다. 테라스를 넓게 뽑아내자 더 큰 방이 만들어졌고 조형적인 형태가 집 전체의 틀을 이루면서 마당까지 만들어냈다.

15m² - 25m²

Slice

The contrasting colors of the black exterior against the white interior helps giving the impression of the building being 'a slice' in the garden.

The fundamentals for this house are: it has to be a lovely place to spend an afternoon, spend a night, and a a good place to start the day. Stretching out the terrace resulted in a bigger room and a sculptural shape that frames the whole house while creating a yard.

"A house doesn't need to be more than 15m² to be called a house." As a consequence of the global economic recession and increase of centralization, architects worldwide take on the challenge of designing small houses. The design ideas are often proportionally inverted by the sizes.

A small building on just 15m² is just as likely to give the architect a pleasurable task than a ginormous building on 5000m². A significant reduction of size and budget is not necessarily synonymous with a reduced ingenuity. Designing small houses even grant a greater sense of freedom. They are ideal for young architects to start of their careers with.

'How can we get the most out of this project?', was the question Saunders and his colleagues had in the back of their minds when they first sat down to draw the extra and outdoor guest room that the client wanted. They worked with the premises given by the site, and that meant preserving the old, existing plum trees in the garden. As a result the plum trees now grow through the triangular building, undisturbed. The contrasting colors of the black exterior against the white interior helps giving the impression of the building being 'a slice' in the garden.

Artists sometimes work on the architects' court, and sometimes architects work in the intersection between art and architecture. Saunders believe that it's easier to let the fantasy run loose on small buildings. At least the chance of succeeding is greater. Artists such as Doug Aitken and Olafur Eliasson have too made small houses. Perhaps the small houses with their sculptural shapes are in an area bordering between art and architecture?

'Slice' describes the little house Todd Saunders drew in a small garden in Slåttevik just outside Haugesund in Norway. The challenge was: how do we design a 15m² sized house in the most exciting way possible? The buliding was supposed to be small, yet have a perfect design.

The fundamentals for this house are: it has to be a lovely place to spend an afternoon, spend a night, and a a good place to start the day. In the end, 'Slice' became more than what we usually expect from a house, and also more unusual.

Stretching out the terrace resulted in a bigger room and a sculptural shape that frames the whole house while creating a yard.

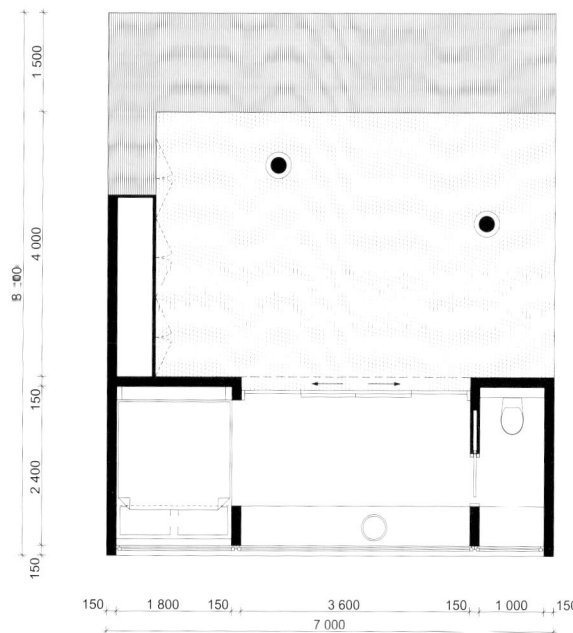

PLAN

21.2m²

Prefabricated Movable

Location
Movable

Use
Living, short term stay(hotel), office, study (classroom), retail

Built Area
21.2m²(1F), 5.2m²(Mezzanine)

Total Floor Area
26.4m²(+3.5m²(Outdoor terrace))

Floor
1 floor + mezzanine

Exterior Finish
Concrete

Interior Finish
Cross laminated timber (softwood)

Project Architect
Ülar Mark

Construction
Kodasema OÜ

Photographer
Tõnu Tunnel, Oliver Moosus, Paul Kuimet

Mini house KODA
사전제작하여 배송하는 이동식 주거, 코다

Kodasema

15m² - 25m²

모듈은 다양한 용도에 맞춰 변경할 수 있기 때문에, 코다주택은 주거공간, 단기 체류호텔, 사무실, 상점, 강의실 등으로 활용되고 있다. 이러한 이동식 주택을 결합하면 2층짜리 주거단지도 만들 수 있다.

KODA는 이동식 콘크리트 조립 주택이다.
KODA는 디자인했다기 보다는 개발했다는 표현이 어울리는 주택으로, 여러 분야 간에 이루어진 협력의 산물이자 현재 진행형의 협력 과정이기도 하다. 비교적 길었던 개발 과정 덕분에 모든 측면의 세부사항까지 꼼꼼히 검토할 수 있었다. 이 주택이 담고 있는 주요 콘셉트는 지속가능성, 건강한 실내 기후, 혁신적인 생산 과정과 자재 이용, 스마트 기술과 시간을 초월한 간결한 공간이 이루는 조화 등이다.

자연 채광을 최대한 활용한 설계와 스마트 난방, 환기, 공기 정화 시스템, 빌트인 자동제어 시스템 덕분에 에너지 낭비를 최소화할 수 있었다. 단열 효과가 좋은 벽과 거대한 4중창 파사드로 열 손실을 최소화했고, VIP(진공단열패널)을 이용한 유리 파사드의 열관류율은 0.1 W/m²/K 및 0.3 W/m²/K이었다. 또한, 태양전지판을 설치해서 매년 사용하는 양보다 더 많은 에너지를 생산하는 플러스 에너지(plus-energy) 주택을 만들었다.

공장에서 생산된 178mm 두께의 얇은 복합 패널은 외부는 콘크리트, 내부는 목재로 제작했으며 그 사이에 두께가 60mm밖에 되지 않는 시즈티뷔패닉을 삽입했다. 이러한 자재의 조합으로 견고하고 내구성 있는 외관과 자연스럽고 안락한 분위기의 실내 공간을 만들어졌고, 필요에 따라 건물 냉난방이 가능하다. 외부 표면에 별노의 마감처리를 하지 않아, 구조를 이루는 재료가 그대로 마감재 역할을 한다. 견고한 구조와 건물 토대 역할을 하는 바닥패널 덕분에 땅을 파서 기초공사를 할 필요도 없다.

모듈은 다양한 용도에 맞추어 변경할 수 있는데, 현재까지 KODA 주택은 주거공간, 단기 숙소(호텔), 사무실, 상점, 강의실 등으로 활용되고 있다. 콘크리트 외벽이 있어 시골 호숫가에서부터 붐비는 도심에 이르기까지 모든 장소에 KODA 주택을 건축할 수 있다. 이동식 주택을 결합해 2층짜리 주거단지도 만들 수 있다(층수는 콘크리트 구조의 지지력을 보강하여 늘릴 수 있다). 토대가 없는 이동식 주거단지는 현재 개발 대기 중으로 향후 5-15년간 비어 있었을 이곳 도심 부지에서 임시 주거지 역할을 담당하고 있다.

1 5.2m² SLEEPING AREA FOR 2 ON THE MEZZANINE
2 3.1M2 SHOWER AND TOILET
3 OPEN CLOSET
4 FULLY EQUIPPED KITCHEN
5 16m² LIVING ROOM CEILING HEIGHT 3.5M
6 3.5m² TERRACE
7 SOLAR PANELS
8 CONCRETE EXTERIOR
9 NATURAL WOOD INTERIOR
10 NO FOUNDATION

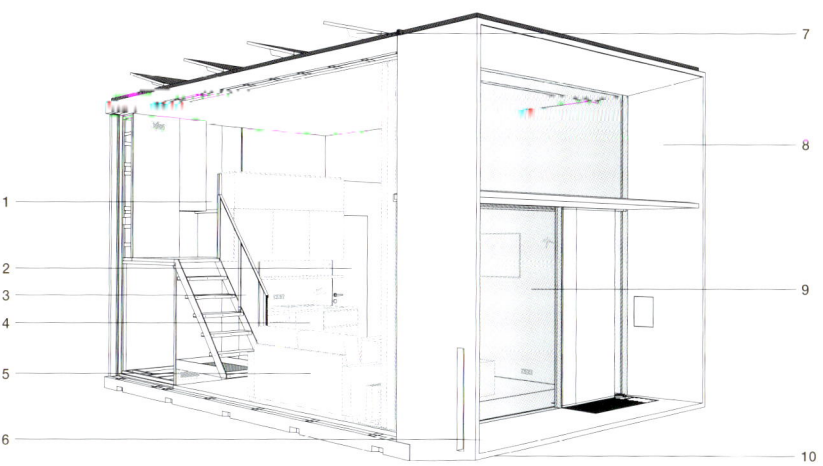

AXONOMETRIC

15m² - 25m²

The module can be adjusted for various uses, so far KODA houses have been used for living, short term stay hotel, office, retail and classrooms. Combining the movable houses up to two story villages can be built.

KODA is a prefabricated movable concrete house.
KODA is a house that has rather been developed than designed. It is a result and an ongoing process of teamwork between different disciplines. The relatively long development process has allowed to consider every detail in all the aspects. Sustainability, healthy indoor climate, innovation in production method and material use and smart technologies meeting timeless quality of simple space are the main concepts combined in this house.
Energy waste is minimal thanks to a design that makes the most of natural sunlight, smart heating, ventilating, and air conditioning system, and a built-in automation system. The heat transferred through the well-insulated walls and the large four-ply glass facade is minimal, with VIP (vacuum insulated panels) solution the U-value of the walls is 0.1 $W/m^2/K$ and 0.3 $W/m^2/K$ for the glass facade. The solar panel solution available ables the house to return more power to the grid on an annual basis than it uses making it a plus-energy house.

The factory-made 178mm thin composite panels are made of a concrete exterior and wood interior, with only 60mm vacuum panel insulation in-between. This combination of materials creates a strong and durable exterior, a natural and warm interior and will keep the building as warm or cool as needed. In addition there is no need for extra finishing layers - the structural materials are the finishes. Thanks to its solid structure and a floor panel that also acts as a plate foundation there is no need for digging up the ground and laying foundations.

The module can be adjusted for various uses, so far KODA houses have been used for living, short term stay (hotel), office, retail and classrooms. Thanks to the concrete exterior the house can be placed to all imaginable locations from a lakeside rural area to the dense city centers. Combining the movable houses up to two story villages can be built (this can be easily increased by strengthening the load bearing concrete constructions). The foundation free movable villages act as a temporary refill on city centre plots that otherwise are vacant for the next 5-15 years and and waiting to be developed.

$15m^2 - 25m^2$

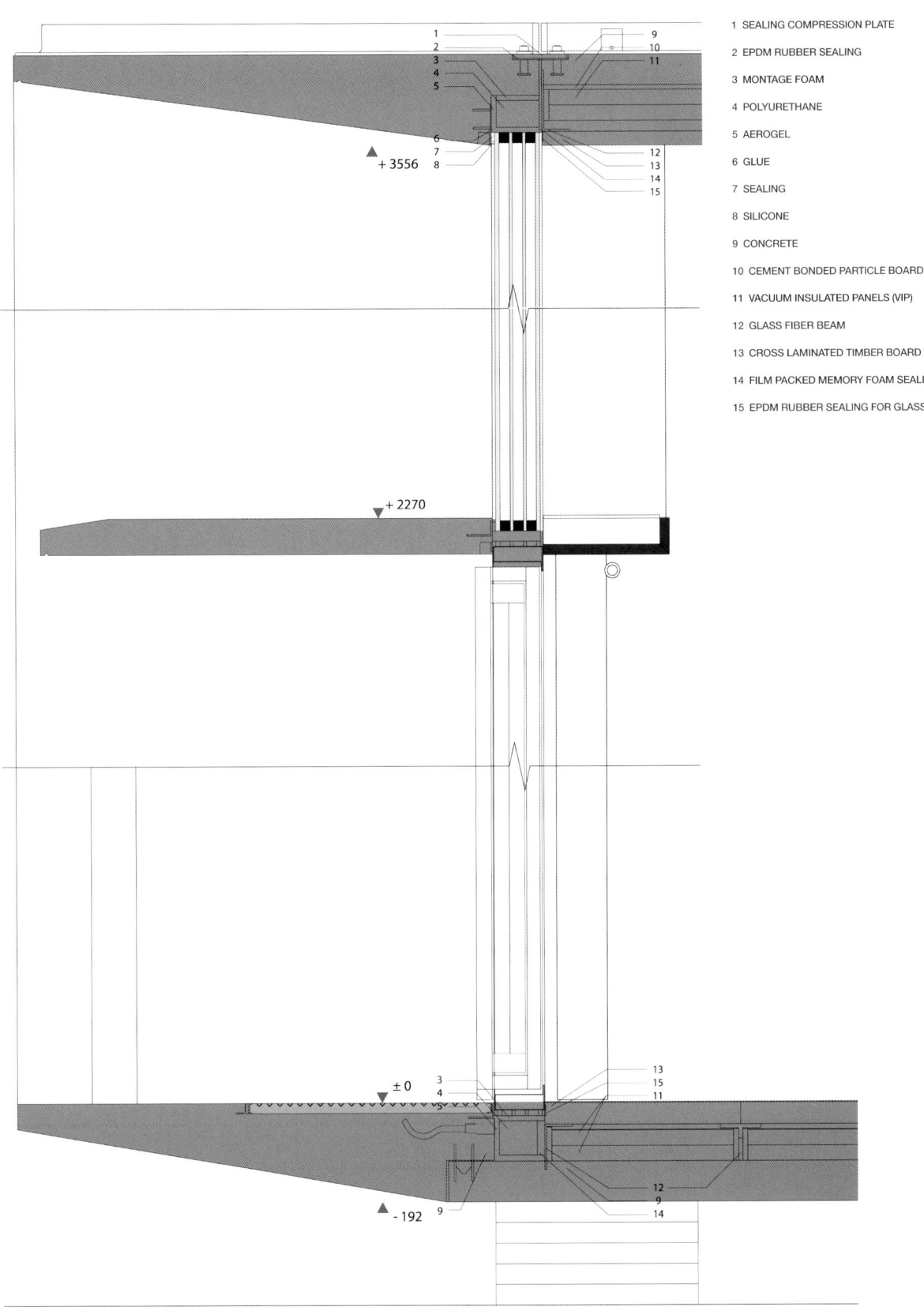

1 SEALING COMPRESSION PLATE
2 EPDM RUBBER SEALING
3 MONTAGE FOAM
4 POLYURETHANE
5 AEROGEL
6 GLUE
7 SEALING
8 SILICONE
9 CONCRETE
10 CEMENT BONDED PARTICLE BOARD
11 VACUUM INSULATED PANELS (VIP)
12 GLASS FIBER BEAM
13 CROSS LAMINATED TIMBER BOARD (CLT)
14 FILM PACKED MEMORY FOAM SEALING
15 EPDM RUBBER SEALING FOR GLASS

DETAIL SECTION

KODA

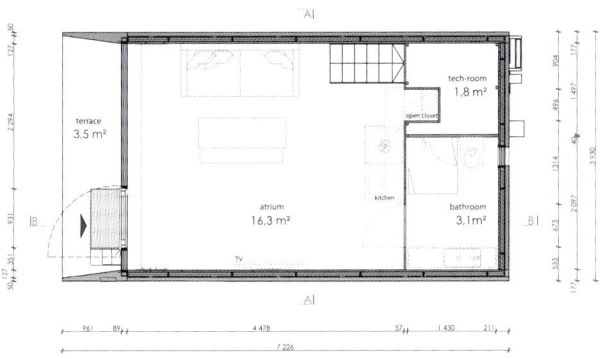
1ST FLOOR PLAN

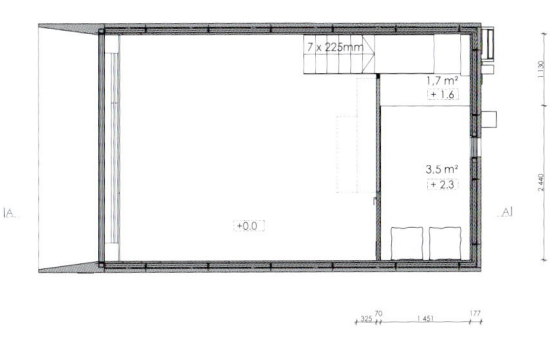
2ND FLOOR PLAN

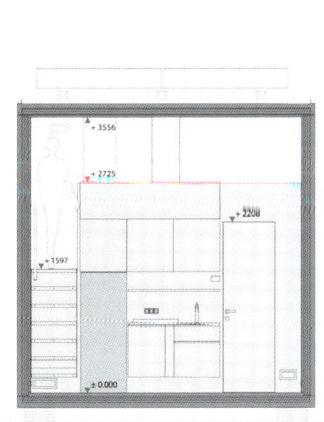
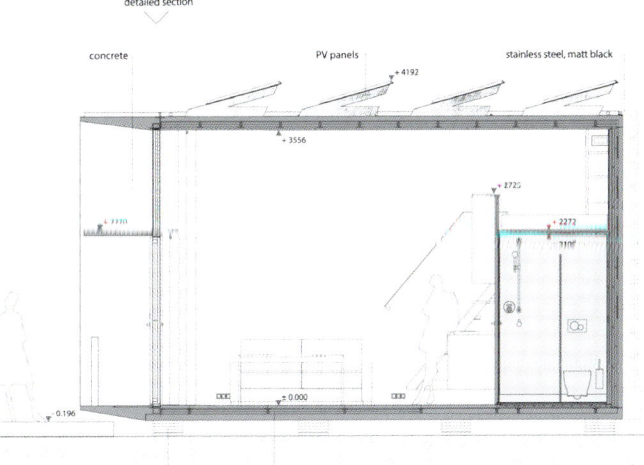
SECTION

Kodasema

interview

최소의 공간을 극대화하는 전략은 'NO DESIGN'이다.

일반 콘크리트보다 3배 더 견고한 벽체를 사용하여 벽두께를 축소할 수 있었다. 코다 주택은 토대가 필요 없는 모듈로써 시설이 완비된 채 배송된다.

Q. 모듈의 표준 크기를 어떻게 계획하였는가?

A. 설계 초기 과정의 목표는 마음껏 이동할 자유를 제공하고 공간 안에서 해방감을 느낄 수 있는 주택을 짓는 것이었다. 3.5m 높이의 천장과 전체가 유리인 파사드를 통해 사적인 공간이 모듈 외부로 연장될 수 있도록 하고, 16㎡ 규모의 주거공간에서 매우 넓은 공간감을 누릴 수 있도록 설계했다. 공간을 최대한 자유롭고 유동적으로 활용할 수 있도록 내부에는 가구를 따로 고정하지 않았다. 거주인은 공간 안에서 태극권, 요가, 스트레칭 또는 왈츠를 추는 등 원하는 활동을 자유롭게 할 수 있어야 한다. 주택의 최종 규격은 1:1 모형 제작과 인적 실험을 통해 결정했다.

개별 모듈은 주거에 필요한 모든 기능을 갖추고 있다. 주방, 화장실, 수면 공간, 거실 모두 표준 규격에 맞춰 설계했으며 표준 설비를 이용할 수 있어 최종 이용자에게 더 많은 자유를 안겨준다. 북유럽식 공간 활용을 기준으로 하면 25㎡ 규모의 KODA 주택은 1.3명이 거주하기에 적합하다.

Q. 최소한의 공간을 극대화하는 아이디어는 무엇인가?

A. 최소 공간은 불필요한 요소를 모두 제거함으로써 극대화할 수 있다. KODA 미니 주택을 개발할 때 총괄 전략은 '노 디자인(No design)!'이었다. 설계 과정에서 불필요한 요소를 없애기 위해 많은 것을 우리 스스로 만들어냈다. 예를 들어 KODA에는 창문이나 문틀, 마룻장이 없으며 벽, 바닥판, 천장판은 마감 처리를 하지 않아서(외벽은 내후성 콘크리트 처리가 되어있고 내벽에는 아름다운 직교적층 목재를 이용했다) 별도의 마감재를 덧댈 필요가 없다.

KODA

디자인을 단순하고 미니멀하게 유지하면 그 온전한 공간을 마음껏 사용할 수 있어 거주인의 자유가 가능해진다. 기술 설비(난방, 냉방, 환기, 자동제어, 수도, 하수 및 기타 통신 설비 등)는 자동차 엔진처럼 주택의 뒤편에 모두 배치했다. 이런 방법을 통해 발전 및 노후 속도가 빠른 시설을 손쉽게 제거하고 교체할 수 있다.

Q. 예산을 줄이거나 또는 제한된 예산 내에서 최대한의 품질을 내기 위한 전략은 무엇인가?
A. 자동차나 전자제품 등의 예와 비교했을 때 건축 산업에는 변화가 거의 없다. 우리는 바로 이 점에 변화를 주어, 주택을 자동차처럼 제작하기로 했다. 부지 내 건축 공사가 공해(먼지, 소음)를 유발하고, 긴 시간이 소요되며, 위험하고 정밀성이 떨어지는 반면, 우리는 고품질의 패널을 빠르고 정확하게 생산할 수 있는 조건을 갖춘 공장에서 패널을 생산해낸다. 또한, 새로운 생산 방식을 도입하기도 쉽다. 이는 빌딩싱이 있고 납땜한 숫자라면 어디든 무릎 생산 공장을 손쉽게 세울 수도 있다는 것을 의미한다. 거푸집은 도면을 공장으로 보내 합판으로 제작했다. 이후 현지 인력과 자재를 활용해 주택을 생산했다.

우리의 디자인 주요 전략인 '노 디자인' 그리고 불필요한 모든 요소의 제거는 비용, 자재 외에 각종 요소에도 적용된다. 우리가 사용하는 콘크리트는 일반 콘크리트보다 세 배 더 견고해 벽체의 두께를 178mm까지 줄일 수 있었으며, 주택 전체에 들어가는 콘크리트의 양은 $9m^3$밖에 되지 않는데 이는 '일반' 주택의 기초 공사에 쓰이는 양보다 적다.

KODA는 토대가 없는 녹립적인 건물이기 때문에 땅에 구멍을 파고 콘크리트로 메우는 데 드는 비용을 다른 곳에 투자할 수 있다. 토대가 필요 없는 모듈은 시설이 완비된 새 배송될 수도 있다. 이러한 조건 덕분에 지표면에 집을 세우고 통신 장비를 연결하는 설치 시간이 4-7시간밖에 들지 않는다. 전체 모듈과 필요한 각종 요소가 공장에서 생산되기 때문에 공사에 드는 작업과 인건비도 극적으로 줄일 수 있다.

이 프로젝트의 설계 과정과 책임 관계는 일반적인 기존의 고객-건축가-건설자 환경에서 이루어지는 것과 완전히 달랐다. 코다세마(Kodasema)가 개발 과정에서 설계자, 생산자, 고객의 역할을 모두 담당했다. 이런 조건을 바탕으로 디자인의 혁신과 자유를 확보할 수 있었고 일관성이 부여됐다. 우리 회사에는 과거 버전의 모듈과 각 현장을 통해 축적된 지식이 있으며 이를 다음 프로젝트에 활용할 수 있다. 결론적으로 설계 일관성, 연속성 및 대량생산은 한 채의 주택을 짓는데 소요되는 작업 시간을 최적화하고 동시에 디자인 품질과 이에 담기는 사고의 수준을 높이는 방법이 된다.

Kodasema

interview

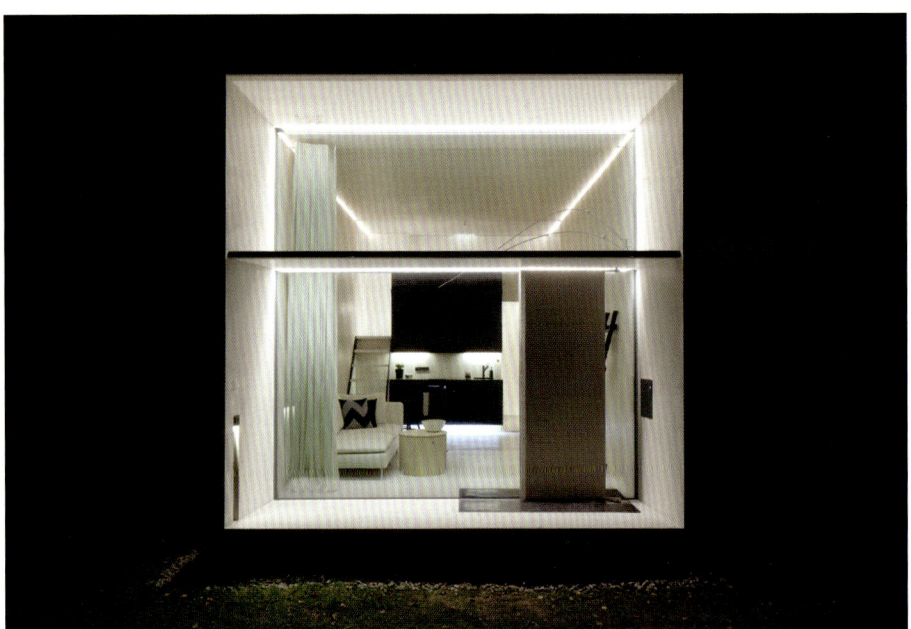

Q. How did you plan for the standard size of the module?
A. From the beginning of the design process the freedom to move freely and feel free inside the house has been an objective. The 3.5m ceiling height combined with a full glass facade that allows the personal space to extend outside the module give the 16m^2 living space very spacious feeling. No furniture is necessarily fixed to this space to keep it free and as flexible as possible. One should be able to freely do tai chi, yoga, stretching or dance valz in this space. The final measurements of the house have been the result of a 1:1 model sketching and personal testing.

The unit has all necessities for living. The kitchen, wet room, sleeping area and living room are designed considering standard sizes and is suitable for standard equipment - this again allows more freedom for the end users. Considering the Nordic use of space, 25m^2 KODA is perfect for "1.3 dwellers".

Q. How can minimum space be maximized?
A. Minimum space can be maximized by removing all redundant/excessive. "No design!" has been our design strategy while developing every aspect of KODA mini house. Throughout the design process we have invented a lot by

ourselves to remove everything redundant/excessive. For example KODA has no window or door frames, no floorboards, the wall, floor and ceiling panel need no finishing (weather proof concrete on the exterior and aesthetical cross laminated timber on the interior), meaning there is no need for extra layers or finishing works.

Keeping the design simple and minimalistic allows the pure space to be universal and leaves the user more freedom. We have decided to keep all the technical solutions (heating-cooling-ventilating system, automatisation, water, sewerage and other communications) all together at the back of the house, similar to a car engine. This way these systems that both develop and age the fastest can be easily removed and replaced.

Q. Within a limited construction budget, would you be able to share your tips to make a higher quality project?
A. Looking at the building industry we see there has been very little change compared to the production of cars or electrical devices for example. We decided to change that and produce houses like cars. Instead of building on site which is polluting (dust, noise), time consuming, dangerous, imprecise, etc. we produce the panels in a factory where the environment is supportive to produce fast and precise high quality panels. In addition the new production method can be easily reproduced, meaning that the factory of the houses can be easily built wherever necessary and reasonable. The drawings of the formwork are sent to a workshop where the formwork is cut out of plywood. Local workers and materials are then used to produce the house.

Or main design strategy 'no design' and losing all excessive is also acute when talking about the cost and consumption of materials and elements. The concrete we use is up to 3 times stronger than regular concrete, allowing to minimize the overall wall thickness to 178mm - at the same time using only $9m^3$ concrete for the whole house which is less than is needed for a foundation of a 'normal' house.

KODA is a freestanding and needs no foundation therefore the expenses on digging a hole and filling it with concrete can be invested elsewhere.

Without a need for a foundation the module can be shipped fully equipped. This results in a 4-7 hour installation time on site where the house is placed on a level surface and connected to the communications. The prices of construction works and cost on skilled workforce can be dramatically reduced by preparing the elements or whole modules in a factory.

The design process and responsibility has been totally different from the standard/traditional client-architect-builder situation. During the development process Kodasema has been the designer, producer and the client all in one. This has allowed for innovation and freedom in design combined with consistency. In the company we have accumulated the knowledge gained from each version and site and are able to pass it on to the next projects. At the end of the day the design consistency/continuity and mass production is a way to optimize the work hours spent on one house and at the same time raise the quality and thought put into the design.

HOUSE IN KAWASAKI 100p

YEONNAM-DONG MIX-USE HOUSING 112p

HOUSE IN THE CITY 124p

TRANS 140p

H33167 152p

NEST 160p

[CREVICE] 1740 170p

SMALL HOUSE WITH FLOATING TREEHOUSE 184p

SIXTEEN ROOMS 194p

Mrs. FAN'S PLUGIN HOUSE 206p

SEONGSAN-DONG MIX-USE HOUSING 218p

ACUTE HOUSE 232p

NARVARTE TERRACE 246p

25m² - 35m²

34.8m²

High ceiling
Louver
Folding door

Location
Kanagawa, Japan
Use
House
Site Area
81.2m²
Built Area
34.8m²
Total Floor Area
96.3m²

Floor
B1 - 3F
Structure
Steel flame

Construction
Kudo Komuten
Photographer
Daici Ano / Taichi Mitsuya & Associates

House in Kawasaki

대형 접이식 문을 이용하여 실내외를 연결한 가와사키 주택

Taichi Mitsuya & Associates

$25m^2 - 35m^2$

북쪽에는 자갈길 바닥에서 천장까지 9미터 높이의 큰 방을 하나 만들었고 남쪽에는 4개 층에 걸쳐 개인 공간을 배치했다. 넓게 트인 공간은 상단 채광과 함께 중간에 루버(louver) 바닥을 시공해 두 개의 공간으로 자연스럽게 나뉘도록 했다. 위층에는 큰 창문을, 아래층에는 골목길을 향해 접이식 문을 설치했다.

이 주택은 도쿄 교외의 한 중층 주거 밀집 지역에 자리하고 있다. 고객은 두 아이를 둔 40대 부부로, 근처에 사는 친척 십여 명을 한 번에 초대할 수 있는 넓게 트인 실내 공간과, 차량 두 대를 세울 수 있는 주차 면적도 필요로 했다. 대지 면적은 81.2m²였다.
건물 북쪽은 골목길과 접해 있고 남쪽은 이웃집과 맞닿아 있다. 북측 골목은 막다른 자갈길이라 지나가는 차가 많지 않다. 이 골목에 거주하는 주민 대부분은 길의 일부인 집 앞 공간을 마치 자신의 정원처럼 사용하고 있었다. 우리는 이런 독특한 상황을 활용해보기로 했다.
집 앞 골목을 정원이나 작은 공원처럼 생각하고 이 맥락에 주택이 잘 어울릴 수 있도록 하기 위해 여러 방안을 생각해 보았다.

오랜 고민과 친구 끝에 북쪽에는 자갈길 바닥에서 천장까지 9미터 높이의 큰 방을 하나 만들었고 남쪽에는 4개 층에 걸쳐 개인 공간을 배치했다. 넓게 트인 공간은 상단 채광과 함께 중간에 루버(louver) 바닥을 시공해 두 개의 공간으로 자연스럽게 나뉘도록 했다. 위층에는 큰 창문을, 아래층에는 골목길을 향해 접이식 문을 설치했다.
이렇게 설계함으로써 개인 공간과 자갈길 모두 볕이 들고 바람이 통할 수 있었다. 집주인뿐 아니라 이웃에게도 밝고 편안한 길이 되도록 하고자 했다.
접이식 문을 활짝 열면 실내 공간이 마치 베란다처럼 사람들을 반긴다. 동시에 자갈길은 실내 공간의 연장처럼 느껴진다. 초록 화분으로 장식한 루버 바닥을 통해 실내외 모두에 밝은 채광과 통풍이 이루어진다.
이 집에서 개인 중심의 닫힌 공간은 좀 더 열린 형태의 준개방 공간으로 연결되고, 이는 다시 완전히 열린 골목이라는 개방 공간으로 이어진다. 이러한 단계적 변화를 기획한 이유는 가족의 생활을 대지경계선 안쪽으로만 제한하지 않고 주변 주택가의 풍경 안으로 녹아들 수 있게 하고자 함이었다.

House in Kawasaki

We created a one big room with 9m ceiling height between the gravel path on the north and the four layers of private rooms on the south. This large open space has a top light and it is softly divided into two spaces by the louver floor. The upper part has a huge window and the lower part has the folding door toward the street.

This house is located in a dense medium-rise residential area in the suburbs of Tokyo.
The client is a couple of 40s with two kids. What they requested for the house was a big open space where they can invite their relatives up to ten people who live nearby. Also they expected to have a parking area for two cars. The plot is 81.2m²
It faces on the street on the north while the southern side it faces toward the neighbor's buildings. This street on the north is a dead end passage with gravels and there are not so many traffics. The most of neighbor arrange the frontal area which is the part of the street as like their own garden. That was quite unique situation and it inspired us.

With this context we started from considering this street as a garden or a park and tried to figure out how the building can relate to it.
As a result of repeated studies we created a one big room with 9m ceiling height between the gravel path on the north and the four layers of private rooms on the south. This large open space has a top light and it is softly divided into two spaces by the louver floor. The upper part has a huge window and the lower part has the folding door toward the street.

Therefore both the private rooms and the gravel path can get the sunlight and the wind throughout this space. By handling this we aimed for making this path bright and comfortable not only for the client but also for the neighbors.
When the folding door is completely open the interior space become like a veranda which welcomes people. At same time the gravel path can be recognized as an extension of the interior. And the louver floor with green pots bring brightness and fresh richness to the both interior and exterior.
In this house the private areas with the human scale continue to the large scaled semi-public area and then it continues to the public area (the street) which has the city scale.
Making this gradation we intended to avoid limiting the family's life by the property line and tried to merge their life to the scenery of the residential area.

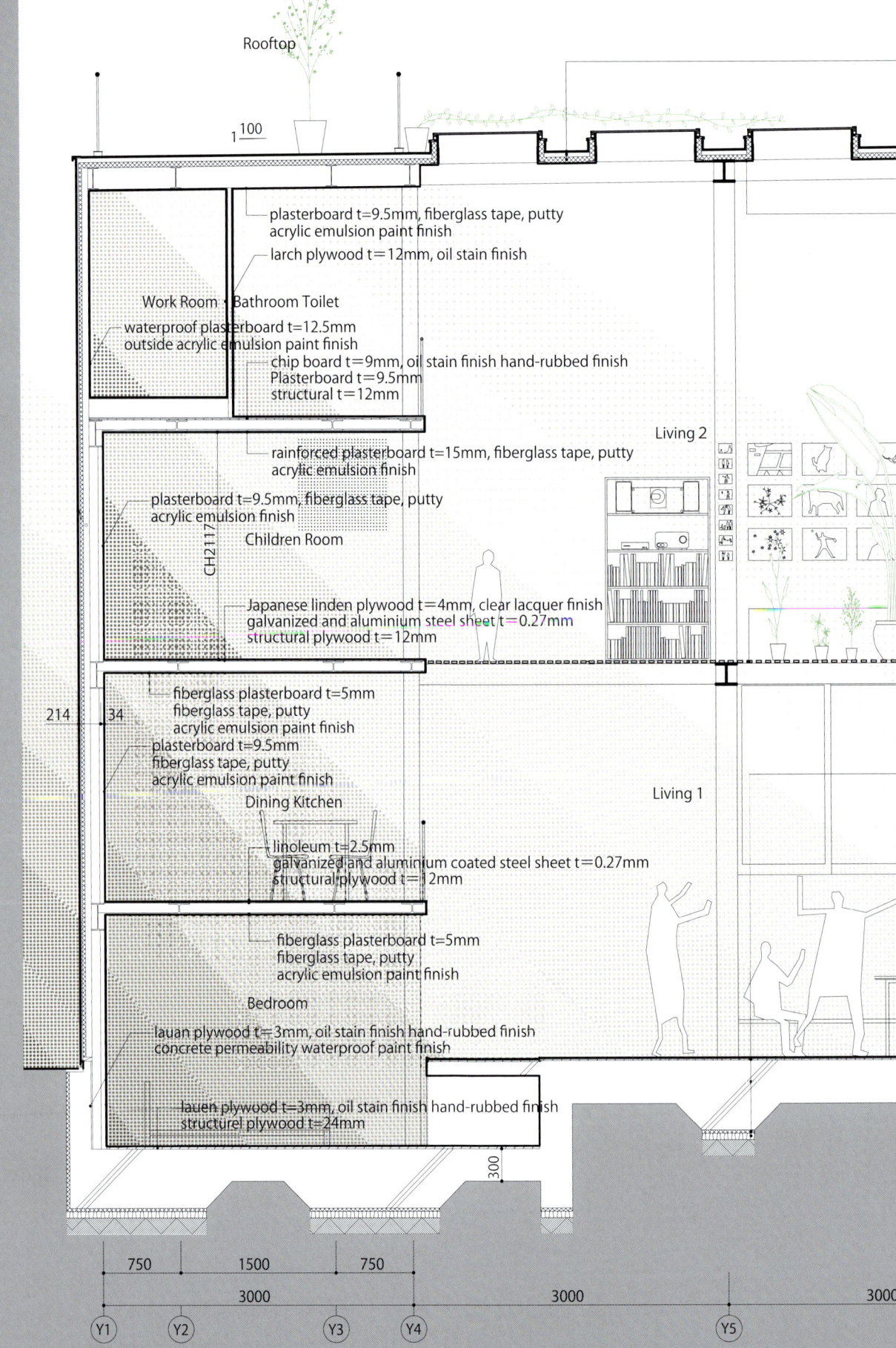

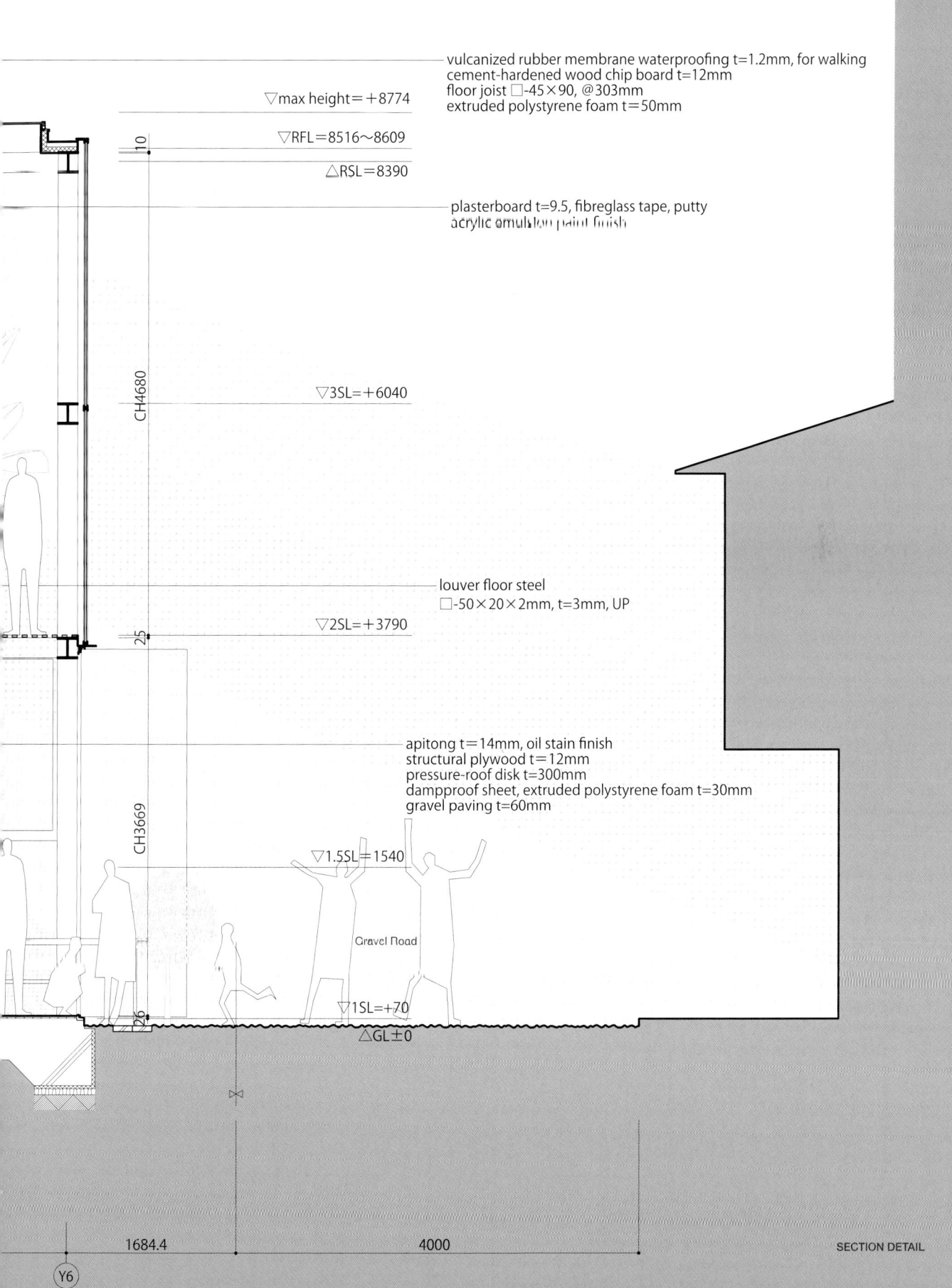

$25m^2$ - $35m^2$

House in Kawasaki

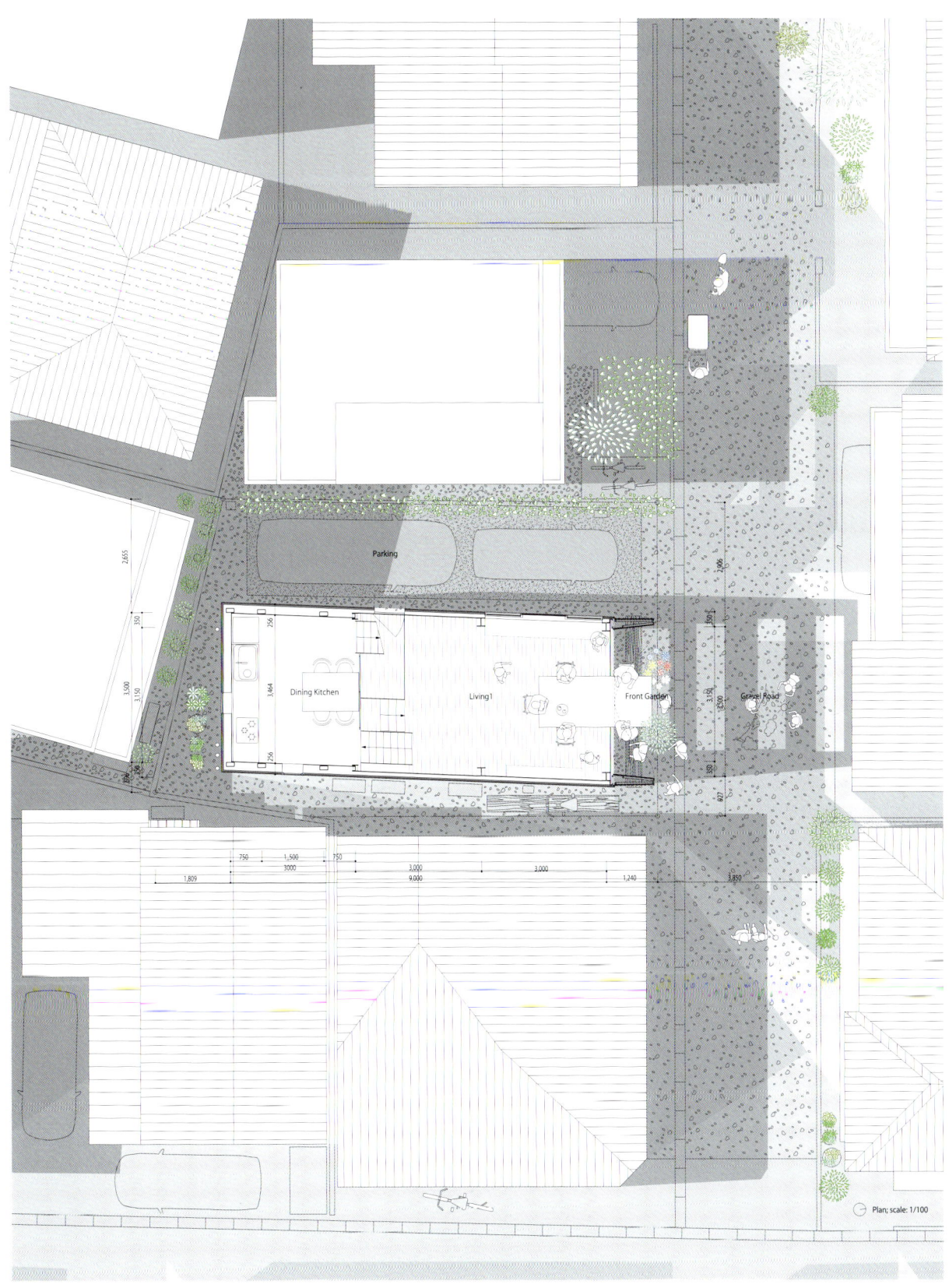

PLAN

Taichi Mitsuya & Associates

interview

**집의 전면인 북측은 골목을 향하고 반대편은 건물로
둘러싸여 있어 채광이 이루어지기 어려웠다.
이를 극복하기 위해 건물 상단 하나의 면으로 된
개방형 지붕으로 모든 방에 햇빛이 들어올 수 있도록 했다.**

Q. 클라이언트의 요구사항은 무엇이었으며, 그에 따른 건축적 대응은 어떤 것이었나?
A. 클라이언트의 요구사항 중 하나는 친척 열 명 정도가 모여 홈파티를 즐길 수 있는 공간이었으면 한다는 것이었다. 그래서 1층에 접근이 용이한 넓은 생활 공간을 고안했다. 이 공간은 골목길과 연결된다.
또한 그들은 밝은 집을 원했다. 집의 전면인 북측은 골목을 향하고 반대편은 건물로 둘러싸여 있어 채광이 이루어지기 어려웠다. 이를 극복하기 위해 건물 상단 하나의 면으로 된 개방형 지붕으로 모든 방에 햇빛이 들어올 수 있도록 했다.

Q. 협소한 대지 위, 수납공간을 창출하기 위한 아이디어는 어떤 것이었으며, 수직적으로 프로그램 공간구성을 어떻게 하였는가?
A. 집은 3.5m×9m 규모의 박스 형태였다. 9m 중 6m는 생활 공간에 할애하고 나머지 3m에 작은 사적 공간을 배치했다.
생활 공간은 골목 방향으로 두 개 층에 걸쳐 있고 반대편에 작은 방들이 있다.
작은 방들은 협소하지만(3.5m×3m) 사적인 기능과 각종 설비를 갖추었다.

이 방들은 특히 위·아래층 거실로 모두 연결되기 때문에 생각보다 훨씬 넓게 느껴진다.

Q. 예산을 줄이거나, 제한된 예산 안에서 최대한의 퀄리티를 만들어내기 위한 방법은 무엇이었나?
A. 요구사항이나 발생하는 문제가 매번 다르기 때문에 모든 프로젝트가 각기 새로운 도전이다. 하지만 어떤 상황에서든 세부적인 문제를 해결하면 이는 곧 비용 절약으로 이어진다.
대부분의 문제는 부분적으로 대응해서 해결할 수 있고 이는 내 건축 방식에서 중요한 기준이다.

House in kawasaki

Q. What were your client's requirements and your architectural approach for them?
A. One of the client's requirement was to have a space what can gather about 10 people who is relatives for home parties. So, I thought in a wide living space which could be easily accessible on the ground floor. This space is in relationship with the alley (street).
On the other hand they wanted a bright house. As the front of the site = north side is facing the street and the other side, is surrounded by buildings it was very difficult to bring daylight. To this, I planned a building under one-roof and provided overhead opening in the roof in order to spread the light to all rooms.

Q. What were your ideas to utilize the space & storage on small site and how to zone the program in a vertical way?
A. I made a box of 3.5m × 9 m. I allocated 6 m of 9 m to the living space and 3 m to the small space. Living spaces are stacked on the street side and small rooms on the other side. Small rooms are very compact spaces (3.5m×3m), they contain private and utility functions.
Moreover small rooms connect both living rooms so the feeling in these spaces are much more spacious than we could expect.

Q. Within a limited construction budget, would you be able to share your tips to make a higher quality project?
A. Each project is a new challenge because demands and problems are changing all the time. However solving details is very helpful to avoid « spending money », to do this, mostly problems were solved by working with sections what are extremely important in my architecture.

29.58m²

Flexibility

Location
Seoul, South Korea
Use
Housing &
neighbourhood facility
Site Area
55.66m²
Built Area
29.58m²
Total Floor Area
103.65m²

Floor
B1 - 3F
Structure
Reinforced concrete
Exterior Finish
STO

Project Architect
Park Changhyun
Photographer
Chin Hyosook

Yeonnam-dong Mix-use Housing
다양한 공간감을 가진 연남동 주택

a round Architects (에이라운드 건축)

$25m^2 - 35m^2$

Yeonnam-dong mix-use housing

> 주택 내부 형태의 다양한 공간감은 내부를 더욱 풍성하게 만들어주고 각 공간마다의 성격을 강화시켜준다.

도심 속 협소주택

서울의 밀도는 점차 높아지고 큰 필지에 높은 건물뿐 아니라 작은 필지의 건물도 점점 늘어나고 있다. 오래된 연남동 주택가에 1960년대에 지어진 단층 주택의 자리에 새로운 계획을 시작하였다. 양평에서 평수가 큰 아파트에서 생활해왔던 젊은 부부는 아이를 위해 서울로 거주지를 옮기면서 작은 골목에 접해 있는 20여 평의 대지를 구입하였다. 앞 도로의 폭이 좁아 대지의 일부를 도로로 내어주고 남은 대지의 크기는 16평 남짓. 작은 골목이 북측에 위치한 대지는 남북으로 긴 형태를 띄고 있었고 대지의 형태고 직각이 아니기에 건폐율을 채우기는 어려웠다. 좁은 골목의 길이가 짧아 앞쪽 6m 도로에서는 건물이 보일 수 있는 가까운 거리이기에 통행이 많은 6m 도로에서 건물이 잘 보일 수 있고 호기심을 유발할 수 있도록 디자인 이뤘고 반대 대지에 지어진 1층은 외대를 위한 근린생활시설이기에 외부에서 건물의 인상이 중요하기도 하여 반대로 골목에서 먼 쪽으로 입구를 계획하고 가까운 쪽은 건물의 첫인상을 만들 수 있도록 계획하였다.

근린생활시설과 주택

한 건물에 지하와 1층에는 근린생활시설이 2층부터 다락까지는 세 식구가 사는 단독주택이 함께 구성되었다. 근린생활시설과 주택과의 관계에서 1층 도로와 접한 외부 공간을 매개로 지하와 1층, 2층을 연결시켜주고 도로에서는 프라이버시를 위해 측면에 입구를 계획하였다. 골목 정면에서는 건물이 오브제처럼 보이지만 측면의 입구를 통해 진입하면 작은 외부공간이 캐노피, 우체함, 공용화상실, 계단으로 구성되어 각 부분을 연결시켜준다. 지하는 남측으로 빛을 늘일 수 있도록 작은 썬큰을 두어 환기도 함께 해결하였고, 지하 외부 계단은 2층으로부터 내려오는 자연 빛이 지하 아래까지 유입된다. 1층의 외부공간은 결국 근린생활시설과 주택을 이어주는 역할을 한다.

협소주택

1층 외부공간을 거쳐 2층 현관으로 올라가는 외부 계단은 다시 골목 쪽을 바라볼 수 있게 크게 열려 있다. 2층 올라가는 계단에서는 골목과의 관계를 다시 연결시켜주는 역할을 하고 이곳을 통해 시선, 바람, 빛의 변화를 경험하게 된다. 2층에 다다르면 주택의 현관 도어를 만난다. 그 도어는 주택과의 첫만남이고 따뜻한 질감의 티크 원목과 섬세하게 제작된 도어의 손잡이를 잡게 된다. 이 손잡이는 가죽으로 감싸 주택의 첫인상을 따뜻하게 맞이하는 첫인상이 된다. 무게감이 느껴지는 두터운 현관 도어를 열고 들어가면 남측에 큰 창을 통해 들어오는 외부 풍경과 빛으로 공간의 크기가 극대화 된다. 북측 어두운 외부 공간과 두터운 도어는 내부로 들어가면서 내부의 공간이 남측 창을 통해 밝고 긴 하나의 공간이 외부와 연결되면서 공간의 크기를 조절해 준다. 작은 주방은 기능적으로 필요한 내용과 수납으로 구성되어 있고, 계단의 하부는 창고로 거실의 가구는 최소한으로 해 거실의 크기를 넉넉하게 느낄 수 있도록 구성하였다.

3층으로 올라가는 내부계단은 이 건물에서 가장 공간적으로 좁고 높은 공간의 형태를 띄고 있다. 주택 내부 형태의 다양한 공간감은 내부를 더욱 풍성하게 만들어주고 각 공간마다의 성격을 강화시켜준다. 3층의 구성은 침실과 딸 방, 그리고 화장실로 구성되어 있다. 딸 방의 도어는 1.5m 폭의 슬라이딩 도어이다. 외부 창과 연결된 복도가 좁아 평소에는 딸 방 도어를 열어 하나의 공간으로 사용할 수 있도록 벽 같은 도어를 만들어 계획하였다. 나중에 딸이 들어오게 되면 실제로 사용할 수 있도록 복도와 방을 하나로 확장할 수 있도록 하였다. 마지막으로 위층은 다락과 외부 테라스를 구성하여 외부 테라스에서의 다양한 기능을 고려하였다. 협소주택은 면적이 적다는 특징이 있어 이름이 붙었지만 이곳은 주택이다. 삶에 대한 방식은 공간과 크기에 많은 영향을 미칠 수 밖에 없다. 삶을 변화시키고 그것을 함께 만들어 나가는 것이 협소주택의 첫 출발이듯이 주택에 대한 방향과 그곳에서 생활 할 사람에 대한 고민이 농밀하게 되지 않으면 어려운 작업이 될 수 있다는 것을 느낀다.

Lifestyles inevitably influence space and scale. As changing life and building it together is the first starting point of a small house, the process would not be possible without setting the clear direction of the house and dense deliberation towards the resident.

Small Houses in Urban Space

Density of Seoul has grown higher over time. Not only skyscrapers in large plots but also small buildings in narrow plots have increased. New project took its place on old Yeonnam-dong residential area site with single story house built in 1960's. Young husband and wife, who used to live in large apartment in Yangpyeoung and moved in to Seoul for their child, purchased about 20py-sized lot adjacent to narrow street. The leftover site area was about 16py, after giving away some portion of the lot to the road. The street was located in north side and the site was longer north to south. Furthermore the angle was not perpendicular, which made even harder to meet the maximum building coverage. The length of the narrow street was too short. So the building was visible from the 6m road with a lot of traffic and thus we designed the building to arouse curiosity. Impression of the building was critical since the basement and first floor will be used as commercial space. The entrance was planned in the rear side and the façade fronting the alley was designed to make a positive first impression.

Commercial Facilities and residence

The building consisted of commercial facilities in basement and first floor and housing of three family members from second floor to attic. In terms of relationship between commercial and residence, basement and first, second floor is connected through exterior space bordering the road; the entrance from the road is located on the side for privacy. Seen from the front, the building seem like an objet. Once you walk through the entrance small outdoor space connects the surrounding facilities: canopy, mailbox, public toilet, and stair. Small sunken is provided on the south side for the lighting and ventilation of the basement, and also western sun flow in from second floor to the basement through exterior staircase. Eventually, the outdoor space on first floor becomes a link between commercial and residence.

Small House

Exterior stairs starting from outdoor space on first floor to entrance on second floor is wide open toward the street. The stair re-connects the relationship with the street and one can experience a change in view, wind, and light. When you reach the second floor, you will come across main door of the house. The door is the first encounter with the house, while sensing the warm texture of teak and delicately crafted door handle. The handle is covered with leather, and give the warm first impression of the house. As soon as you pass through the imposing door, the scale of the space is maximized due to the outdoor scenery and sunlight coming through large window on the south. Scale between the dim northern outdoor space and bright long interior space is adjusted by the southern window. Small kitchen is composed of functional requirements and storage, and the space under stairs is used as storage. Number of the furniture in living room is minimized in order to make the space look larger. Staircase to third floor is the most narrow and highest space in the building. Diversity of the spatial scale enrich the interior space and strengthen the characteristic of each space.

Third floor is composed of bedroom, daughter's room, and bathroom. The door of daughter's room is 1.5m wide sliding door. Since the corridor connected to the external window is narrow, wall-like door was designed to use it as a single space when opened. Also in the case when the daughter leaves home, the hallway and room can be expanded into one for study. Lastly the upper floor contains attic and terrace which can provide a wide range of functionality.

Even though the phrase 'small-house' is named after the size, after all this is a home for family. Lifestyles inevitably influence space and scale. As changing life and building it together is the first starting point of a small house, the process would not be possible without setting the clear direction of the house and dense deliberation towards the resident.

SITE PLAN

$25m^2 - 35m^2$

Yeonnam-dong mix-use housing

1 LOFT
2 BED ROOM
3 LIVING ROOM
4 SHOP
5 BATH ROOM

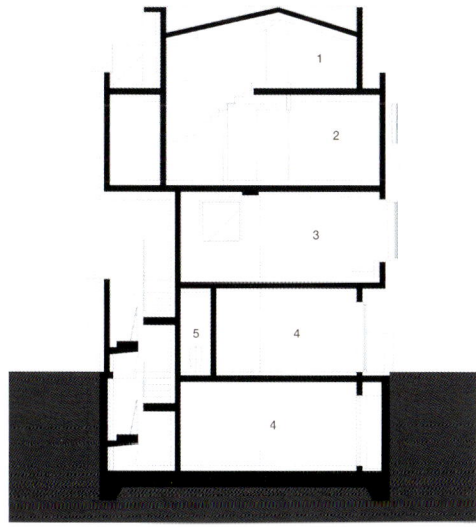

SECTION

$25m^2 - 35m^2$

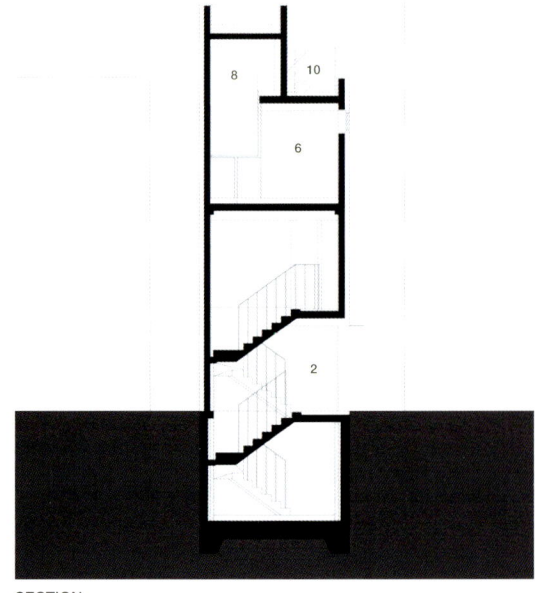

SECTION

Yeonnam-dong mix-use housing

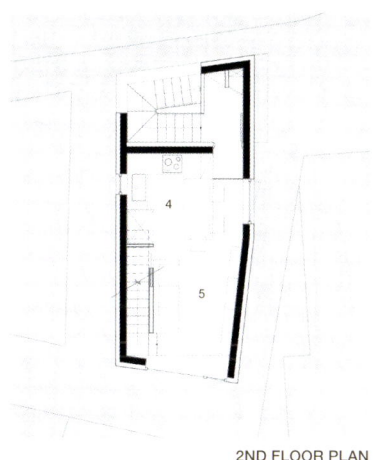
2ND FLOOR PLAN

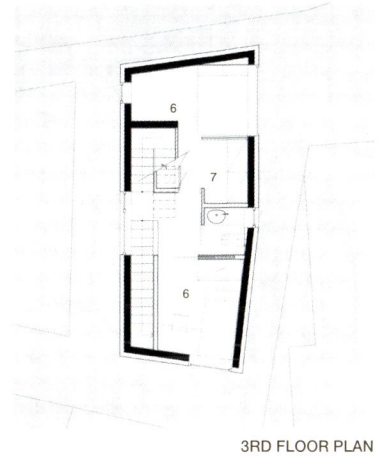
3RD FLOOR PLAN

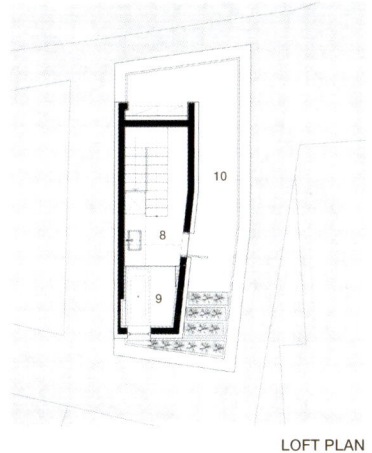
LOFT PLAN

1 SHOP
2 MAIN ENTRANCE
3 SHOP
4 KITCHEN
5 LIVING ROOM
6 BED ROOM
7 DRESS ROOM
8 LOFT
9 BATHROOM
10 TERRACE

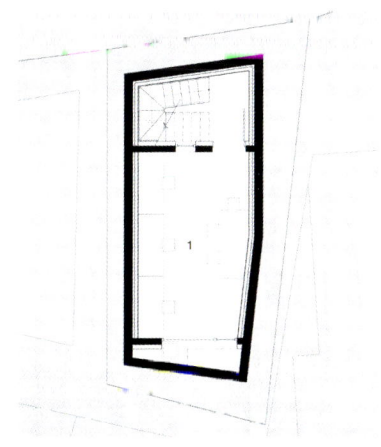
B1 FLOOR PLAN

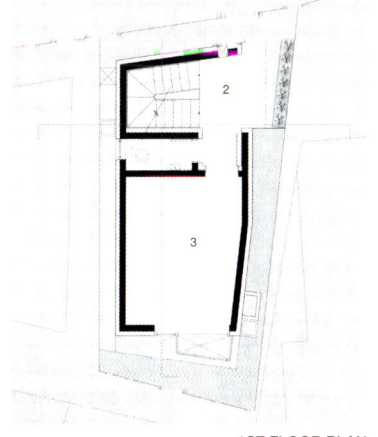
1ST FLOOR PLAN

a round architects, Park Changhyun

interview

**새롭게 지어질 작은 대지의 집에 어떻게 살림들을 넣을 것인가에 대한 질문부터 시작하였다.
살림 중 가지고 갈 것과 처분해야 할 것을 나누는 것부터 시작하여, 가지고 가야 할 것은 새로운 집에 어떻게 배치 할 것인지를 함께 풀어나가면서 집의 계획이 점점 완성되어 갔다.**

Q. 본인이 생각하는 협소주택의 정의는 무엇인가?
A. 협소주택이 물리적으로는 작은 집일 가능성이 높지만 간결한 삶을 바탕으로 정의 내리고 싶다. 대지의 면적이 작아서 그 속에서 삶이 간결하게 되어야 한다기 보다, 번잡하고 펼쳐져 있는 삶을 정리하고 단순하게 사는 장소로써 작은집이라고 생각한다.

Q. 클라이언트의 요구사항은 무엇이었고, 그에 대한 건축적 대응은 어떤 것이었나?
A. 클라이언트는 이전에 서울 외곽의 큰 규모의 집에서 살아 왔었던 분들이다. 이들은 새롭게 지어질 작은 대지의 집에 어떻게 살림들을 넣을 것인가에 대한 질문부터 시작하였다. 그것에 대한 대응으로 기존에 가지고 있던 살림 중 가지고 갈 것과 처분해야 할 것으로 나누면서 이야기가 시작 되었고, 가지고 가야 할 것은 새로운 집에 어떻게 배치 할 것인가를 함께 풀어나가면서 집의 계획이 점점 완성되어 갔다.

Q. 협소한 면적 내에서 최대한의 공간 창출을 위한 아이디어는 무엇인가?
A. 공간에 대한 아이디어는 현재의 삶에서 10년, 20년 후 변화될 가능성을 생각하여 하나의 공간을 나눌 수도 있고 합칠 수도 있도록 계획하였고, 각 실의 용도도 변화될 수 있게 접근하였다. 그리고 좁고 긴 거실에서의 남측 창을 최대한 크게 계획하여 공간의 답답함을 해소하였다.

Q. 일반적으로 협소주택은 주요 동선이 수직적으로 이루어지는 경우가 많다. 이러한 수직적 생활에서 오는 사용자의 불편함을 극복하기 위한 대안은 무엇인가?
A. 수직적인 생활의 불편함은 기본적으로 있을 수밖에 없다. 그렇지만 그 내부에서의 수직 동선에는 다른 곳에서 볼 수 없는 비율의 공간을 계획하여 공간의 풍부함을 제안하였다. 그리고 외부에서는 동네와 연결될 수 있도록 난간의 높이라든지 시각적 오픈을 두어 답답하지 않도록 디자인하였다.

Yeonnam-dong mix-use housing

Q. What do you think "small houses" are?
A. Small houses are highly likely to be narrow physically, but I'd like to define them based on a simple life. I think small houses should be places where one lives a simple life, organizing complicated and uncoordinated life, rather than one's life should be simple in a house because the house is small.

Q. What were your client's requirements and what was your architectural approach for them?
A. The client has lived in a large house in the outskirts of Seoul and asked how the household goods would be put in a new house that was to be built on a small lot first. In response, I suggested dividing the existing household goods into what to bring and what to dispose of first. And as we discussed how the goods would be allocated, and the plan for the house gradually became complete.

Q. What were your ideas for the creation of maximum space within a small site?
A. As for space, the design was performed in such a way that a space could be separated and combined again considering current life and changed life 10 or 20 years later. The use of each room was also approached from the perspective of changes. The southern window was created to be as large as possible to avoid feeling stuffy because the living room was narrow and lengthy.

Q. Generally, the major circulation of small houses is made vertically. What is an alternative to overcome the inconvenience to users caused by this?
A. Discomfort from vertical living basically cannot be avoided. However, abundance of space was suggested by planning the ratio of space which cannot be seen in other places in the vertical circulation inside of the house. And for connection with the neighborhood from the outside, the design was performed without a stuffy feeling with the adjustment of the length of the handrail and visual openness.

27.20m²

Toplight
Rooftop terrace
Intended facade

Location
Tokyo, Japan
Use
House
Site Area
45.84m²
Built Area
27.20m²
Total Floor Area
91.13m²

Floor
B1-3F
Exterior Finish
Resin mortar
Interior Finish
Resin mortar (ceiling), plaster, stone, lauan plywood, larch

Project Architect
Daisuke Ibano, Ryosuke Fujii, Satoshi Numanoi
Construction
Yokota Construction Management
Photographer
GA photographers, Koichi TORIMURA, Yohei Ogata

House in the City
다층구조로 최대 면적을 확보한 도시주택

Daisuke Ibano, Ryosuke Fujii, Satoshi Numanoi

$25m^2 - 35m^2$

House in the city

정해진 대지에서 최대한 넓은 바닥 면적을 확보하기 위해 다층 구조를 도입해야 했다. 층간 수직 이동을 용이케 하기 위해 나선형 구조로 각 층을 분산하고 연결했다. 이 외 복도 없이 방을 서로 연결해 집 내부의 모든 공간이 통로가 되고 층계의 연속처럼 보이는 공간 구조를 설계했다.

이 집은 도심 주택가에 위치해 있다. 부지 주변의 땅은 사찰이 임대해서 쓰고 있어 세대 변화로 인한 구획 세분화는 대부분 모면할 수 있었다. 정원이 있고 회반죽으로 마감한 큰 주택들이 강화 콘크리트와 타일로 지은 견고한 아파트 건물과 나란히 서 있다. 주변 주거시설에 비해 부지가 작았지만 우리는 주어진 환경을 긍정적으로 받아들이고 비교적 보존이 잘된 주택가의 사례를 공유하고 힘고에시 세고운 유형의 노시 수태을 만들어내기로 했다.

우선 정해진 대지에서 최대한 넓은 바닥 면적을 확보하기 위해 다층 구조를 도입해야 했다. 층간 수직 이동을 용이케 하기 위해 나선형 구조로 각 층을 분산하고 연결했다. 이 외 복도 없이 방을 서로 연결해 집 내부의 모든 공간이 통로가 되고 층계의 연속처럼 보이는 공간 구조를 설계했다. 층을 나누면서 집 내부의 각 공간은 면적이 작아졌지만, 계단의 너비를 늘리고 큰 창과 온 가족이 이용할 수 있는 욕조를 설치했다.

또한 옥상 테라스를 두 개 층으로 만들고 모든 요소의 규모를 전체적으로 확장함으로써 활발한 대면과 소통을 도모하고 외부 공동체와의 교류를 촉진할 수 있는 가능성을 부여했다.

부지는 북쪽으로 도로와 맞닿아 있지만, 나머지 3면에는 건물과 접해 있다. 그렇기 때문에 저층부에는 햇볕이나 바람이 거의 들지 않는다. 이 문제를 해결하기 위해 우리는 각 층의 부피를 점진적으로 줄이고 그로 인해 생기는 틈에 천창과 잴루지(jalousie) 창을 시공해 어느 방에서든 하늘로부터 쏟아지는 자연광과 신선한 공기를 즐길 수 있도록 했다. 이러한 창들은 식탁, 침대 겸 소파, 욕조, 벤치 등 다른 요소들과 함께 각 방이 마치 화창한 날 야외에서 휴식하는 듯 안락한 느낌으로 가득 차게 한다.

건물 외벽에는 철골 구조 위로 수지 모르타르 마감이 깔끔하게 처리되어 있고 창틀을 위한 공간이 마련되어 있으며 실내 천장에도 같은 마감이 쓰였다. 이런 디자인은 두꺼운 벽을 강조하며 주변 건물의 견고함을 모방하고 있다. 이와는 대조적으로 실내 공간에는 석재, 타일, 목재 등 다양한 소재가 쓰였다. 우리는 방에 따라 소재를 달리하지 않고 공간상의 위치에 따라 각 소재를 적용했는데 이는 방이 이용자의 행위를 정의하는 현상을 막고 대신 이용자가 직접 여러 공간의 이용 방식을 만들어 내게 하기 위해서였다.

도시에 위치한 주택은 외부 세계로부터 이용자를 보호할 뿐만 아니라 도시적 교류와 소통의 출발점이 되어야 한다. 우리는 이러한 주택의 공공성을 수용하고 안락함을 제공함으로써 주택으로 하여금 거주자의 다양한 실내 활동이 자연스럽게 외부 세계로 발전해 나아가게 하고 이를 바탕으로 개인과 도시가 서로 공감하는 유익한 관계를 형성할 수 있다고 믿는다.

25m² - 35m²

1 BEDROOM
2 PARKING
3 ENTRANCE
4 KITCHEN
5 DINING
6 LIVING
7 TOILET
8 BATH
9 ROOM 1
10 TERRACE 1
11 ROOM2
12 TERRACE 2
13 TERRACE 3

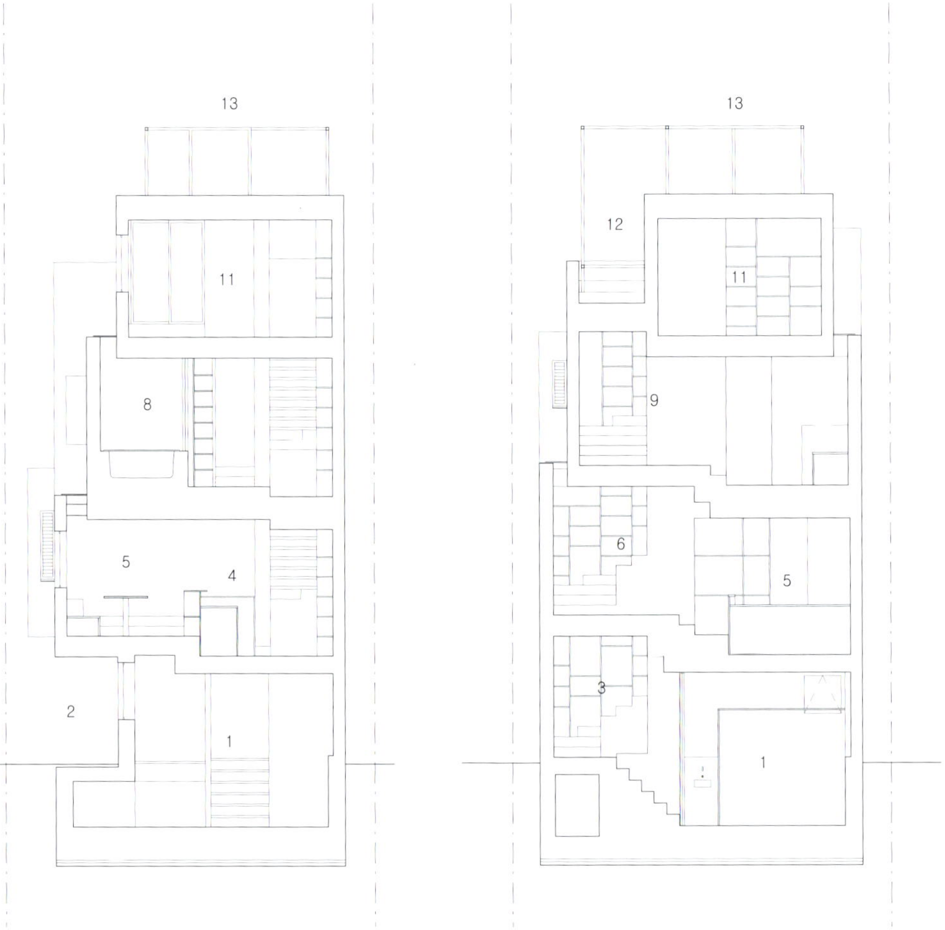

SECTION

A multi-layered structure was necessary to secure a good-sized floor area for this plot. To facilitate vertical movement between these layers, we decided to use a spiral circulation to split and connect the floor levels. We also decided to arrange the rooms adjacently without corridors - thus making every area in the house a pathway, and creating a spatial construction of seemingly successive staircase landings.

This is a house in a centrally located, urban residential area. Since the land around the plot is leased by a temple, it has mostly escaped subdivision caused by generational change. Large mortar-finished homes with gardens stand alongside sturdily built reinforced concrete and tile apartments. The plot is small compared to the surrounding residences, but nonetheless thinking positively about its environs, and sharing and referencing the areas relatively well-preserved residential atmosphere, we set out with the aim to discover a new type of urban home.

Firstly, a multi-layered structure was necessary to secure a good-sized floor area for this plot. To facilitate vertical movement between these layers, we decided to use a spiral circulation to split and connect the floor levels. We also decided to arrange the rooms adjacently without corridors - thus making every area in the house a pathway, and creating a spatial construction of seemingly successive staircase landings. Though the size of individual areas in the house would be reduced due to the split levels, we planned to increase its potential for bringing about encounters and communication, and to create rich connections with the outside community through increasing the width of the stairs, installing large windows and a family-sized bathtub, building 2 levels of rooftop terraces, and generally expanding everything to a larger scale.

The plot faces the road on its northern side, but is bordered by buildings along its other three edges. For this reason, hardly any sunlight nor breeze enter the lower floor. To combat this issue, we incrementally set back the volume of each upper floor, and installed skylights and jalousie windows in the openings so that each room has light pouring in from the sky and enjoys a fresh breeze. Each of these windows are accompanied by other elements such as a dining table, day bed, bathtub and benches etc., so that the rooms are imbued with a comfortable feeling similar to that of relaxing outside on a sunny day.

The outside of the building features a neat resin mortar finish on a steel-frame constructed volume, with spaces for window jambs and running to the indoor ceilings. This serves to accentuate the thick walls, emulating the sturdiness of the surrounding buildings. By contrast, the interior contains various materials including stone, tile and wood. We reiterated these materials according to their position rather than by room in order to encourage occupants to seek out activities in various spaces without allowing the rooms to prescribe their behavior.

As long as residences are located in cities, they not only need to act as protection from the outside world, but also as starting points for urban encounters and communication. We believe that by embracing their public nature and providing comfort, residences allow the diverse internal activities of their occupants to progress naturally into the outside world, creating beneficial relationships in which individuals and cities resonate with each other.

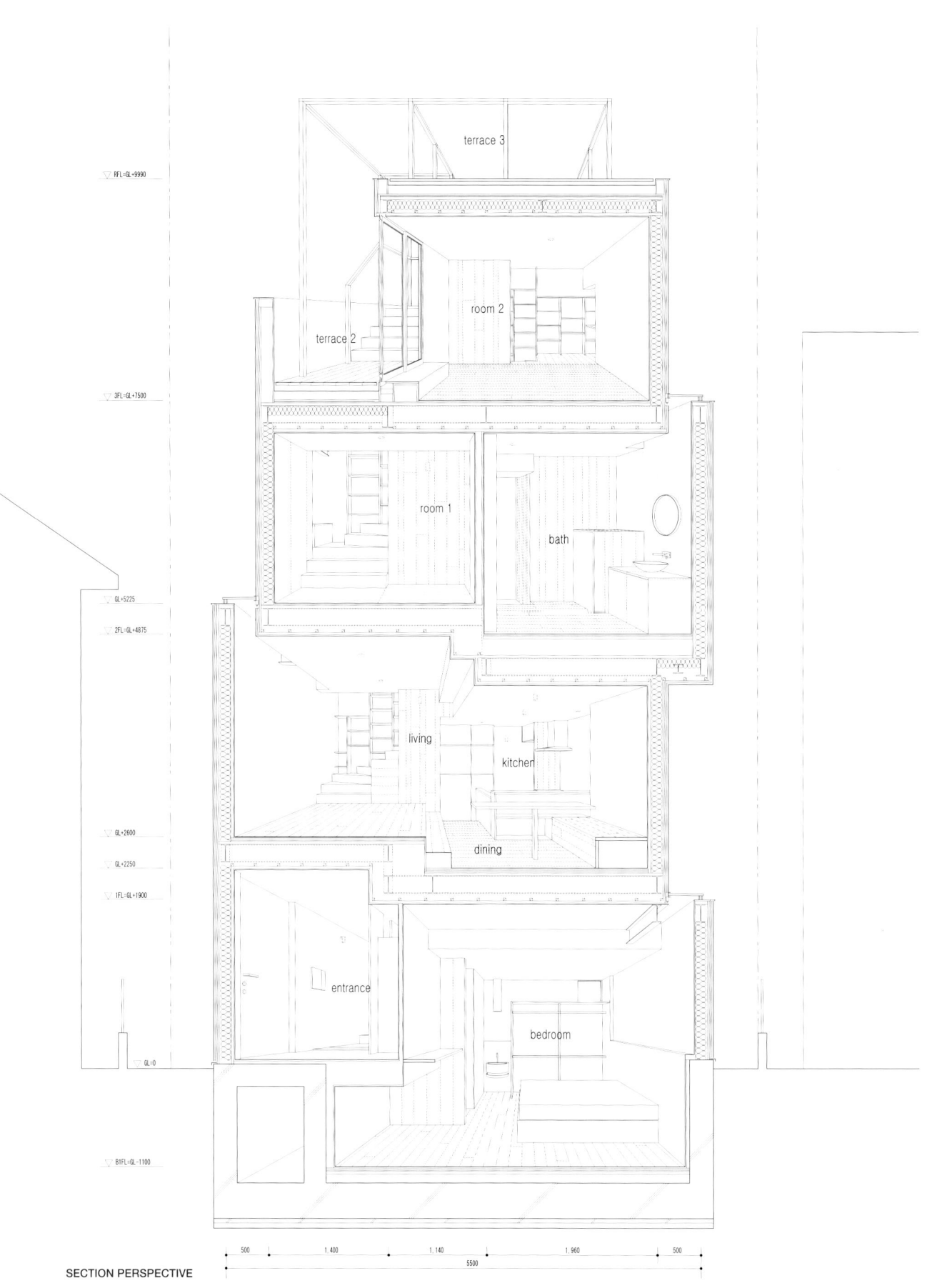

SECTION PERSPECTIVE

House in the city

1 BEDROOM	6 LIVING	11 ROOM 2
2 PARKING	7 TOILET	12 TERRACE 2
3 ENTRANCE	8 BATH	13 TERRACE 3
4 KITCHEN	9 ROOM 1	
5 DINING	10 TERRACE 1	

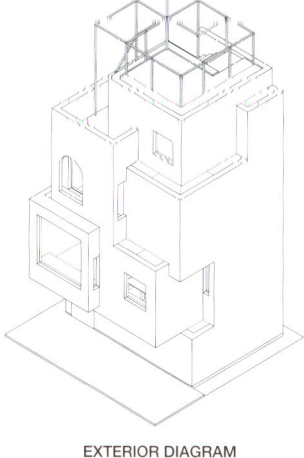

EXTERIOR DIAGRAM

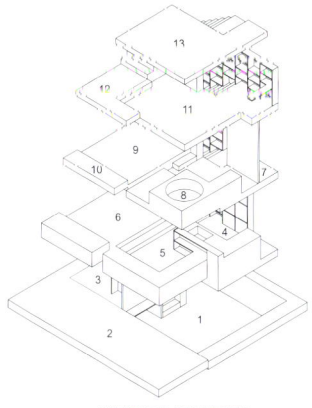

INTERIOR DIAGRAM

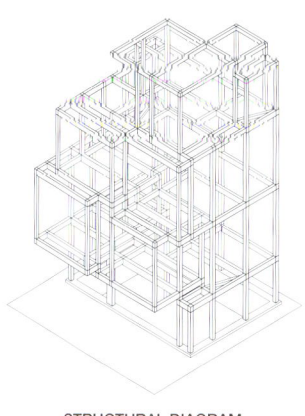

STRUCTURAL DIAGRAM

$25m^2 - 35m^2$

House in the city

1 BEDROOM
2 PARKING
3 ENTRANCE
4 KITCHEN
5 DINING
6 LIVING
7 TOILET
8 BATH
9 ROOM 1
10 TERRACE 1
11 ROOM 2
12 TERRACE 2

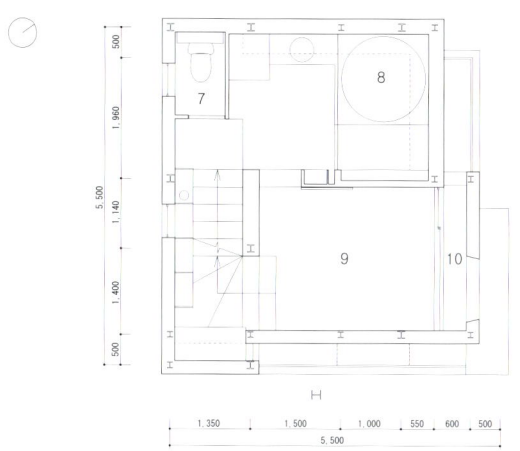

2ND FLOOR PLAN

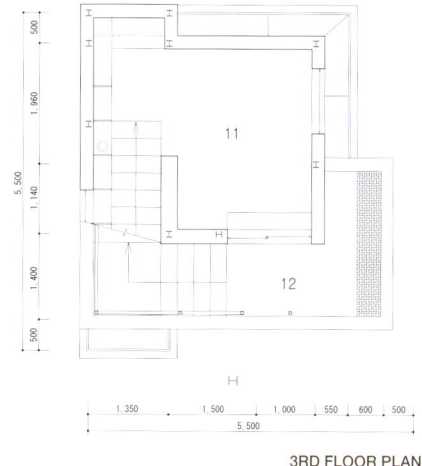

3RD FLOOR PLAN

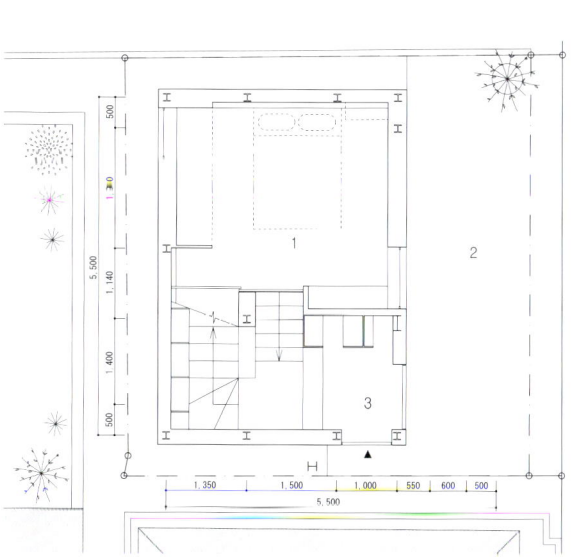

GROUND FLOOR PLAN

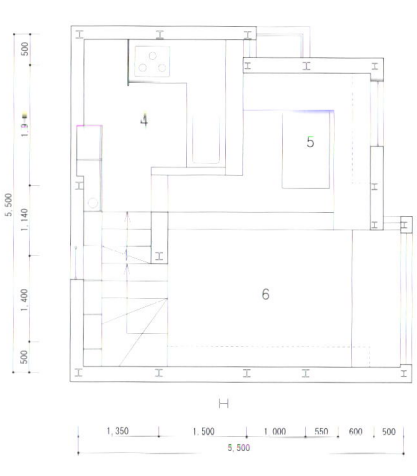

1ST FLOOR PLAN

$25m^2 - 35m^2$

House in the city

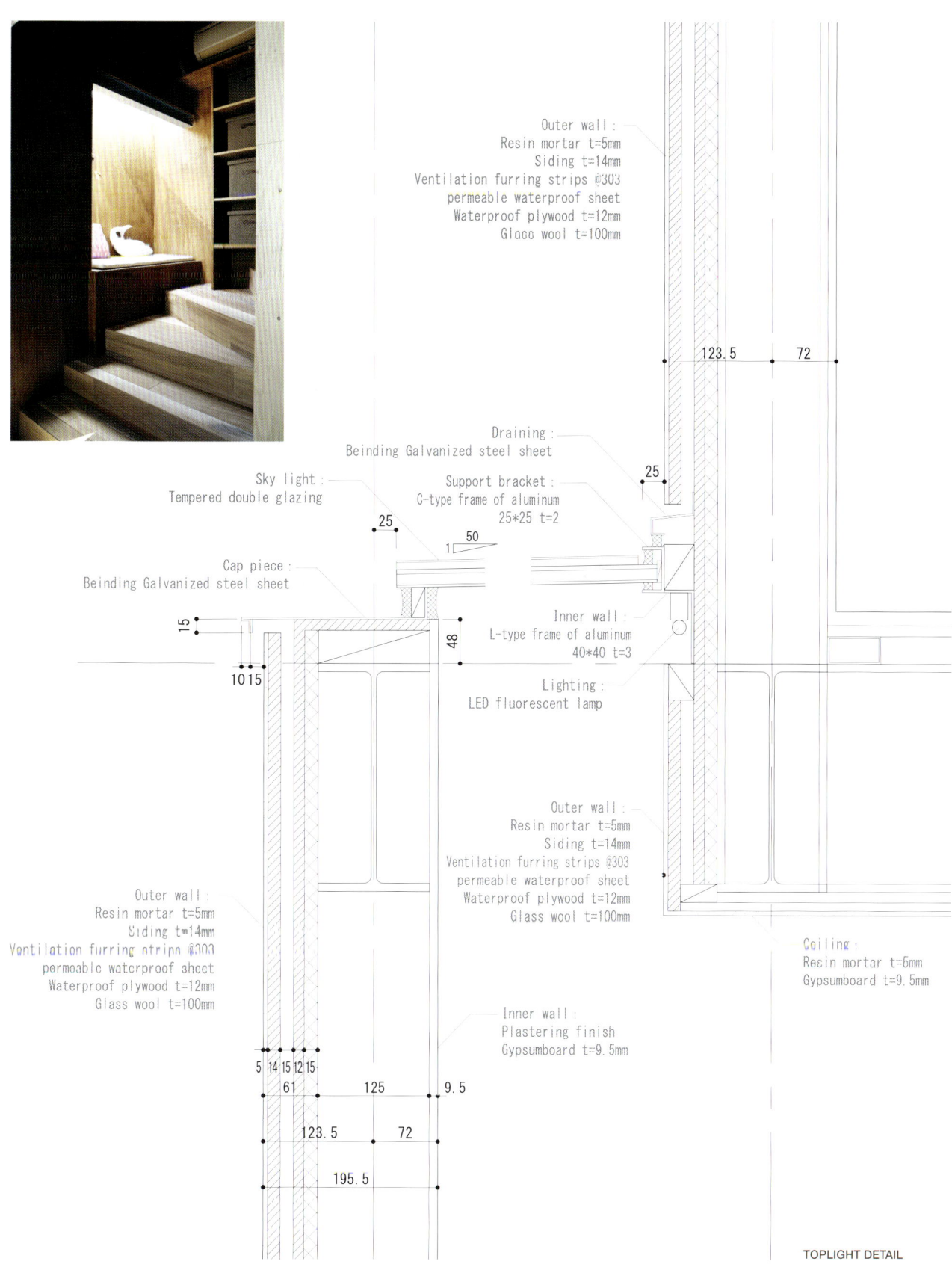

Outer wall :
Resin mortar t=5mm
Siding t=14mm
Ventilation furring strips @303
permeable waterproof sheet
Waterproof plywood t=12mm
Glass wool t=100mm

Draining :
Beinding Galvanized steel sheet

Sky light :
Tempered double glazing

Support bracket :
C-type frame of aluminum
25*25 t=2

Cap piece :
Beinding Galvanized steel sheet

Inner wall :
L-type frame of aluminum
40*40 t=3

Lighting :
LED fluorescent lamp

Outer wall :
Resin mortar t=5mm
Siding t=14mm
Ventilation furring strips @303
permeable waterproof sheet
Waterproof plywood t=12mm
Glass wool t=100mm

Outer wall :
Resin mortar t=5mm
Siding t=14mm
Ventilation furring strips @303
permeable waterproof sheet
Waterproof plywood t=12mm
Glass wool t=100mm

Ceiling :
Resin mortar t=5mm
Gypsumboard t=9.5mm

Inner wall :
Plastering finish
Gypsumboard t=9.5mm

TOPLIGHT DETAIL

$25m^2 - 35m^2$

House in the city

Daisuke Ibano, Ryosuke Fujii, Satoshi Numanoi

interview

나선형 스킵플로어 구조를 통해 연속적인 동선을 설계하였다.

각 용적의 엇갈린 배치로 인해 생기는 틈을 이용해 프라이버시를 보호하면서, 채광을 해결하였다.

Q. 클라이언트의 요구사항은 무엇이었으며, 그에 따른 건축적 대응은 어떤 것이었나?
A. 클라이언트는 우리 건축팀의 일원인 다이스케 이바노(Daisuke Ibano)이다. 이번 프로젝트의 시작은 그가 가족과 살기 위해 도쿄 중심의 작은 주택 부지를 매입하면서 시작되었다(실은 토지 소유주와 일정 기간 동안 임대계약을 맺었다). 기존에 세워진 집은 낡고 좁고 어두웠다. 곧 새 자녀가 태어날 예정이라는 사실을 알게 된 클라이언트는 낡은 집을 철거하고 밝고 살기 좋은 새집을 짓기로 결심했다. 이를 위해 우리는 하나의 건축가 팀을 꾸렸다.

Q. 협소한 대지 위, 수납공간을 창출하기 위한 아이디어는 어떤 것이었으며, 수직적으로 프로그램 공간구성을 어떻게 하였는가?
A. 부지는 그리 넓지 않았고 이 안에서 가족이 생활하기 위해서는 네 개 층이 필요했다. 위아래로 이동하는 방식은 불편하기 때문에 방을 나선형 스킵플로어 구조로 만들어 연속적인 동선을 설계했다. 내구 공간 구성은 건물 외형에 드러난다.
부지에 건물이 인접해 있고 북쪽 도로로만 공간이 트여 있었기 때문에 사생활을 보호하기 위해 북쪽을 제외하고는 외벽에 큰 창을 설계하지 않았다. 그리고 각 용적의 엇갈린 배치로 인해 생기는 위쪽과 측면의 틈을 이용해 채광을 해결하는 방식을 고려했다.
각 용적을 엇갈리게 배치한 또 다른 이유는 법적 규제에 있었다.
일본 주택가에는 맨해튼의 토지사용제한법처럼 후퇴선 제한규정이 있다. 그래서 위층으로 갈수록 건물의 면적을 줄여야 했다.

Q. 예산을 줄이거나, 제한된 예산 안에서 최대한의 퀄리티를 만들어내기 위한 방법은 무엇이었나?
A. 이 건물의 용도는 주택이다. 하지만 나는 이곳을 공공장소를 만들고 싶었다. 왜냐하면 이 주택은 도시에 세워졌고 도시와 이루는 좋은 관계가 집 내부의 관계들로 이어지기 때문이다. 이러한 연유로 나는 모든 방이 문 없이 연결되는 나선형의 연속적인 공간을 만들기로 결심했다. 그리고 다양한 곳에서 비틀린 틈으로 들어오는 바람과 빛을 이용해 시내 도처에 있는 소공원처럼 편안한 장소를 만들었다.
한편 우리는 각기 다른 개성을 지닌 다양한 실내 공간을 창출하고 싶었다. 연속적으로 맞닿아 있는 하나의 공간을 만드는 것도 중요했다. 이렇게 상반되는 요구 사항을 실현하기 위해 모든 방의 바닥과 천장 소재를 통일했으며 반대로 벽은 방마다 다른 마감으로 처리했다. 또한 천장의 마감재와 외벽의 마감재를 통일시켜 방에서도 외부와 연결된 느낌이 들도록 만들었다. 이러한 방법을 통해 정해진 예산 내에서 보다 높은 품질의 결과물을 산출했다고 생각한다.

House in the city

Q. What were your client's requirements and your architectural approach for them?
A. The client is Daisuke Ibano who is also one of our architects team. The beginning of the project is that he first bought a plot with a small housing in the city center of Tokyo to live with his family (actually, he made a contract to borrow this plot with the owner for a certain period). However, the house originally built here was old, small and dark. And when it turned out that a child of the client would be born, he decided to scrap the old house and make a new, bright and livable house. And for that, a team of our architects was formed.

Q. What were your ideas to utilize the space & storage on small site and how to zone the program in a vertical way?
A. The site is not so large, it was necessary to provide four floors to ensure the area required for the family to live. In that case, since the up and down movements are painful, I made it possible to move continuously by making the room a spiral skip floor. The composition of the internal space appears in the external form.
And the fact that the building was near the site and that the only open side was the north road side did not have a large opening on the outer wall surface except north to protect privacy. So we thought about taking in light from the top and side of the staggered arrangement of the volumes.

Another reason for the staggered arrangement of the volume is the legal reason. In the residential area of Japan there is a setback-line limit (like Manhattan zoning law). Therefore, it was necessary to reduce the building volume as going up.

Q. Within a limited construction budget, would you be able to share your tips to make a higher quality project?
A. The purpose of this building is housing, but I wanted to create a public space. Because this house is built in the city and the good relationship with the city leads to the relationship of the inside of the house. For that reason, I decided to have a helical continuous space where all the rooms were connected without a door. And by the wind and light entering from various places by the shifted volume, we arranged a comfortable place like everywhere in the city like a pocket park.
On the other hand, we wanted to achieve a variety of interior spaces with different expressions for each place. And it was also important to create the whole contiguous space. In order to realize such opposing demands, floor and ceiling were made to be unified material in every room, and on the contrary materials of the walls is different for each room. Moreover, by making the material of the ceiling material and the exterior wall material the same, we could feel the connection with the outside even in the room. Through these methods, we are thinking to be a better quality project within the budget.

28.55m²

Void Toplight Terrace

Location
Tokyo, Japan
Use
House
Site Area
42m²
Built Area
28.55m²
Total Floor Area
76.75m²

Floor
3F

Project Architect
*Komada Takeshi,
Komada Yuka*
Photographer
Sobajima Toshihiro

Trans

도시생활을 즐기는
젊은 커플을 위한 주택 트랜스

Komada Architects Office

$25m^2 - 35m^2$

> 이 집은 개인 공간이 모여 하나로 합쳐진다는 개념을 바탕으로 도시와 연결된다. 건물은 규모와 모양이 서로 다른 집들이 모인 작은 마을 같다. 그 안에서 극도로 협소하거나 길고, 높고, 낮고, 최대한 넓거나 최대한 좁은 공간, 그리고 천창과 골목을 모두 하나로 통합하고 있다.

도시에 사는 사람들은 자연의 주기나 정해진 체계에 얽매이기보다 직접 자신에 맞게 시간과 공간을 구획한다. 날짜변경선을 넘나들며 밤과 낮을 건너뛰기도 하고 때로는 자유롭게 거닐 수 있는 도심의 공간을 찾아 헤매기도 한다. 한편 날로 커가는 고층 아파트 시장이 말하듯 장소와의 교감은 더욱 드물어지고, 생활 공간에 활기를 불어넣고자 하는 시도가 늘고 있다.

미나미 아자부에 있는 '트랜스(TRANS)'는 도시 생활을 즐기는 젊은 커플을 위해 설계한 집이다. 트랜스가 있는 이 저지대 구역은 독특한 작업세계를 지닌 장인들이 모여 형성되었다. 시골 마을에서나 볼 수 있는 작은 공장이나 목조 아파트 건물이 여전히 남아있다. 이미 수많은 건물이 신축되거나 재건축되었지만 메이지 시대에 개발된 토지 구획 방식이나 격자식 골목 구조를 그대로 유지해왔기에 이곳만의 특색이 잘 보존되어 있다. 고객은 고층 아파트로 이사를 가지는 않더라도 도시가 주는 이점은 활용할 수 있길 원했다. 대지는 좁은 편으로 입구 쪽의 너비가 4m 정도이다. 이들은 도시 안에서 떠돌지 말고 이곳에 닻을 내리기로 했다. 이들이 도시에서 생활하는 방식은 매력적이고 공감도 됐다.

두 고객은 생활 리듬이 달라 침실뿐만 아니라 나머지 생활 공간도 분리해야 했다. 이 집은 개인 공간이 모여 하나로 합쳐진다는 개념을 바탕으로 도시와 연결된다. 건물은 규모와 모양이 서로 다른 집들이 모인 작은 마을 같다. 그 안에서 극도로 협소하거나 길고, 높고, 낮고, 최대한 넓거나 최대한 좁은 공간, 그리고 천창과 골목을 모두 하나로 통합하고 있다. 지붕에서 보면 마치 협곡을 내려다 보는 것 같고 지층에서 위를 보면 지하에서 올려다보고 있는 듯한 느낌을 받는다. 계단 끝에 놓인 두 개의 방은 다락방 같은 분위기이다. 두 방 사이에 있는 작은 발코니에서 위로 올라가면 옥상이 나오고 그 위에서 도쿄 깊숙한 곳에 서 있는 자신을 발견하게 된다.

The house is a place which connects to the city through the concept of individual spaces coming together as one. The building is like a small town with the variations in scale and shape, of integrated spaces that are extremely narrow, long, high, low, maximum size and minimum size, skylights and an alleyway.

Community Anchor

City dwellers are not bound by natural cycles or old social systems, they control time and space. Day and night are regularly reversed, sometimes rewinding and fast forwarding the time at the International Date Line. At times, it is possible to find several places to "be" in the city, where we can enjoy freely wandering. On the other hand, the connection with the place is quite rare, as symbolized by the growing apartment tower market, accelerating a revitalization of living spaces.

"TRANS" in Minami Azabu is a house designed for a young couple who enjoy city life. This lowland area was formed by artisans who brought their unique works. Some of the small town factories and wooden apartments which were built can still be found today. Although many buildings were new and reconstructed, keeping the style of land allotment and latticed alleys formed in the Meiji Period give a distinctive ambiance which remains in this area. The clients chose not to purchase a tower apartment but to live in a place in which incorporates the strengths of the urban location. The lot is small with an entrance of about 4m wide. Rather than drifting in the city, they dropped an anchor at this place they found. Their style as urban dwelling is attractive and relatable.

The two clients have differing schedules which means they require separate living spaces, aside from the bedroom. The house is a place which connects to the city through the concept of individual spaces coming together as one. The building is like a small town with the variations in scale and shape, of integrated spaces that are extremely narrow, long, high, low, maximum size and minimum size, skylights and an alleyway. From the roof it is like looking down into a valley and from the ground floor like looking up from a basement. Two rooms at the end of the staircases have an attic-like feel. Going up the staircase from the small balcony between the rooms, you will get to the rooftop where you can find yourself immersed deeply in Tokyo.

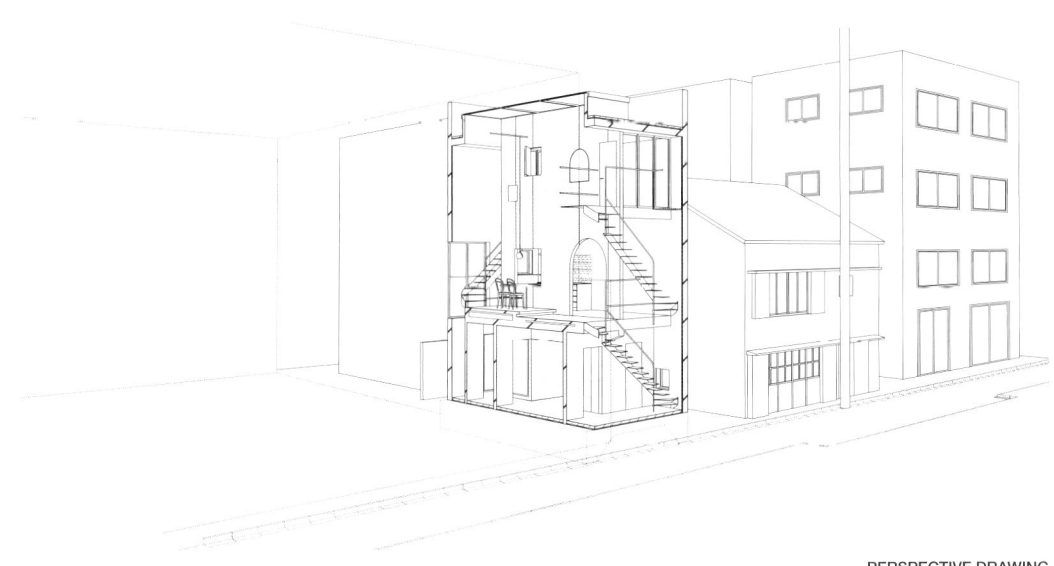

PERSPECTIVE DRAWING

25m² - 35m²

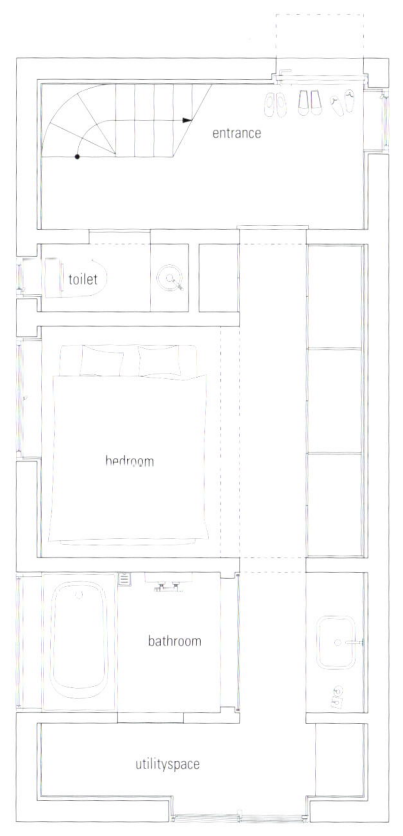
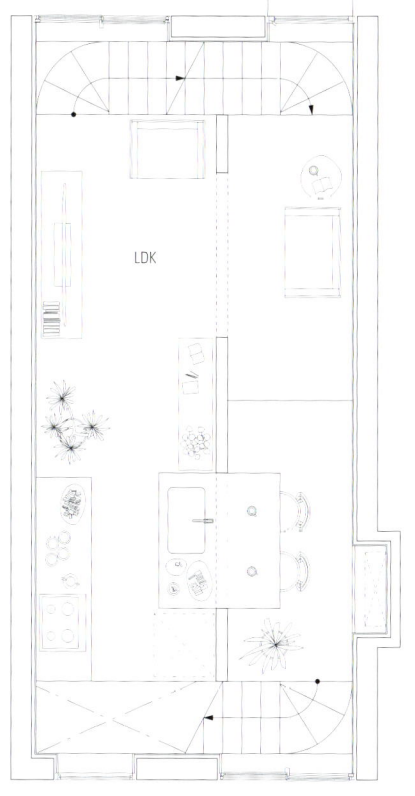

1ST FLOOR PLAN   2ND FLOOR PLAN

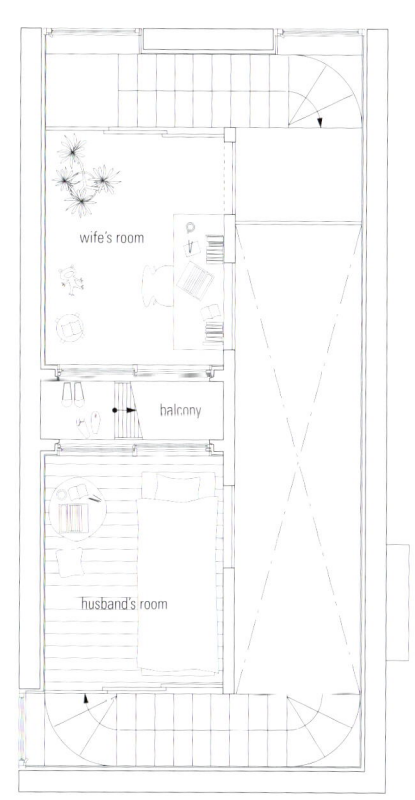

3RD FLOOR PLAN

$25m^2 - 35m^2$

Komada Architects Office

interview

**주거 내부가 2층과 3층을 관통하는
아트리움을 통해 연결되어 다채로운 공간감과
거리감을 느낄 수 있다.**

Q. 클라이언트의 요구사항은 무엇이었으며, 그에 따른 건축적 대응은 어떤 것이었나?
A. 클라이언트는 젊은 커플이었고, 서로의 생활 리듬이 달라 침실 외에도 각자의 개인 방을 필요로 했다. 그래서 우리는 개인 공간을 확보하면서도 이를 부드럽게 연결해 전체적으로 편안한 집을 만들어내고자 했다.

Q. 협소한 대지 위, 수납공간을 창출하기 위한 아이디어는 어떤 것이었으며, 수직적으로 프로그램 공간구성을 어떻게 하였는가?
A. 집의 전면부 너비 중앙 지점에 벽을 세웠고, 이 벽을 따라 필요한 공간을 구획해 나갔다.
최상층의 두 방으로 향하는 각각의 동선을 설계해 1층부터 꼭대기 층까지의 이동이 편안하게 이어지도록 했다. 또한 집 전체가 2층과 3층을 관통하는 아드리움을 통해 언결뇌노록 해 실내에서의 공간감과 거리감이 다채롭게 느껴지도록 했다.

Q. What were your client's requirements and your architectural approach for them
A. The client is a young couple and life rhythm is different, they wanted each individual room apart from the bedroom. So we thought that we create a whole comfortable space by gently connecting them while securing individual rooms.

Q. What were your ideas to utilize the space & storage on small site and how to zone the program in a vertical way?
A. We built a wall at the center of width of front and made necessary rooms along that wall.
Also, an approach was set up in each of the two top floor single rooms, so that the vertical direction from the first floor to the top floor was gently connected using the space. Furthermore, the whole is connected by taking the atrium connecting the 2nd and 3rd floors. It created various scales and sense of distance.

29.19m²

Void Terrace

Location
Seoul, South Korea
Use
House
Site Area
48.67m²
Built Area
29.19m²
Total Floor Area
94.46m²

Floor
4F
Main Structure
Steel reinforced concrete
Exterior Finish
STO

Project Architect
Lee Yongeui, Song Kideok
Construction
KINFOLKS STUDIO
Photographer
Kim Mineun

H33617

불편한 집과 행복한 집 사이 적은집 H33617

GONGGAM&KINFOLKS (공감 건축사사무소)

$25m^2 - 35m^2$

> 한옥의 공간구조를 차용하여
> 각 실들을 '채'로 나누는 구조를
> 계획하고, 이를 연결하는 중간영역을
> 자연공간으로 채워 내부와 외부를
> 연결하도록 하였다.
> 3,4층부터 테라스 공간이
> 등장하는데, 계단을 올라 외부
> 공간(테라스)을 통해 각 방으로
> 들어갈 수 있다. 즉, '집안의 집'이
> 있는 모양이다.

방법과 개념의 차이

積恩集(적은집) H33617는 1.5억의 공사비로 리모델링을 원하던 건축주에게 신축을 제안하며 시작한 프로젝트이다. 작은 땅의 작은 집일지라도 건축주와 자녀들에게 자연의 삶을 느끼게 하고 싶었다. Nisizawa의 Garden&House와 거의 같은 가로x세로의 건물이지만, 정반대의 설계 요소로 같은 개념을 풀어보고 싶었다. 한옥의 공간구조를 차용하여 각 실들을 '채'로 나누는 구조를 계획하고, 이를 연결하는 중간영역을 자연공간으로 채워 내부와 외부를 연결하도록 하였다. 3점식 기둥 대신 벽식 구조로 계획하였고, 벽이 필요하지 않았기 때문에 본래의 기능이 아닌 조경을 위한 요소로 대체시킨 Garden&House보다 더욱더 벽으로 둘러쌓았다.

벽과 벽 사이에 내부와 외부를 연결하는 중간 영역을 계획하였으며, 벽에 생겨난 보이드를 통해 수직적 공간을 연결하고 그 너머를 조망할 수 있도록 계획하였다.

불편한 집과 행복한 집의 사이

積恩集 H33617는 Garden&House와 다르게 4인 가족으로 구성되어 있어 방이 3개 필요하고, 1층에 주차가 가능해야 했다. 2층에는 LDK(거실-다이닝-키친)형의 가족이 모이는 공간을 만들었고, 계단의 난간을 없애 답답함을 최소화했다. 3,4층부터 테라스 공간이 등장하는데, 계단을 올라 외부 공간(테라스)을 통해 각 방으로 들어갈 수 있다. 즉, '집안의 집'이 있는 모양이다.

매우 불편한 집이다. 2층에서 씻고, 3층 또는 4층에 있는 자신의 방으로 갈 때마다 밖을 나가서 들어가야 한다. 비와 눈은 직접 맞지 않게 캐노피를 계획했지만, 비가 오면 비를 바라보며 들어 가아하고, 눈이 오면 몸을 움츠리고 들어가야 한다. 매번 건축주와 미팅 때마다 불편한 집을 설명하며 설득해야만 했고, 도시에서 우리가 어떻게 살아갈 것인가를 같이 고민해주었다.

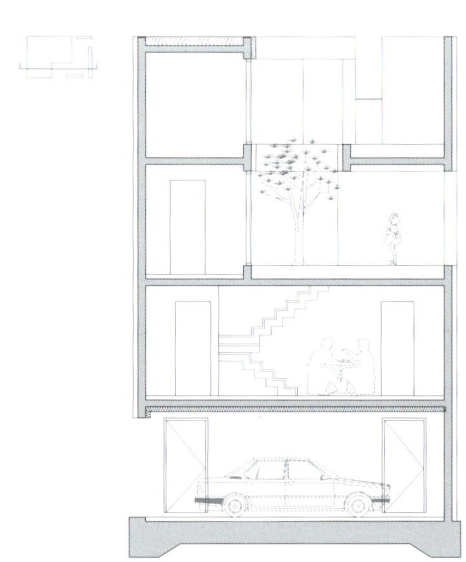

SECTION

Inspired by design and structure of Korean traditional house, Hanoak, individual rooms become the basic units to create a house and the nature is brought into the centralized transitional space that links each room and at the same time it connects the outside and inside. There are terraces on third and fourth floors. To access to the individual rooms on the upper levels inhabitants should pass through the staircase and the terraces which are the external spaces. It would give the occupants a sense as if there is another house inside of the house.

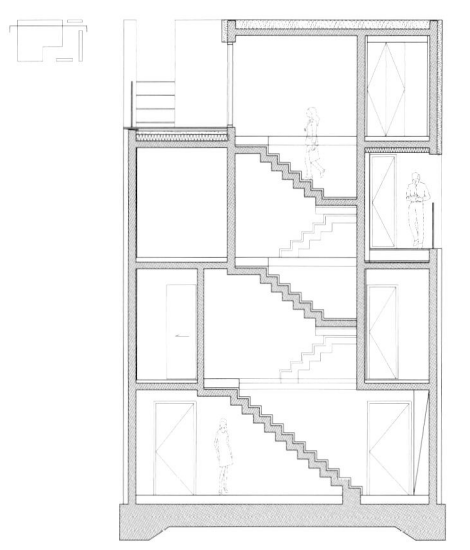

SECTION

Difference between method and concept

H33617 is the project that the client initially wanted home renovation with budget of 1.5 million won. However, we suggested them to go for demolition and a new construction, as we wanted to create a space closer to nature for the clients and their children even in the small house built on a tiny site. While the site has almost the same length of width and depth with Garden & House and shares the same design concept, we wanted to use an opposite design principle as a solution for the project. Inspired by design and structure of Korean traditional house, Hanoak, individual rooms become the basic units to create a house and the nature is brought into the centralized transitional space that links each room and at the same time it connects the outside and inside.

Garden& House chose to use 3-point structure and replaced the wall with landscape element as the function of wall was not needed. However, we emphasize the wall even more for the house of our client.

Conflict between inconvenient space and happy space

Unlike Garden& House, the number of client's family members is four. They asked for 3 private rooms and one parking lot on the ground level. To accommodate these, LDK(Living, Dining and Kitchen) type family gathering space is designed on the second floor, and the balustrades are removed to minimize visual obstacles. There are terraces on third and fourth floors. To access to the individual rooms on the upper levels inhabitants should pass through the staircase and the terraces which are the external spaces. It would give the occupants a sense as if there is another house inside of the house.

It can be a very inconvenient house. Whenever the residents take shower on the second floor, they need to go through the terrace in order to return to their rooms. A canopy was introduced to protect the inhabitants from rain and snow; however, they might still have some inconveniences on rainy or snowy days. Every time when we had a meeting with client we had to persuade the client to understand all the benefits that the design will offer regardless all the inconveniences may the design cause. Also we talked and discussed with them to find a way how they should live well in urban area.

25m² - 35m²

H33617

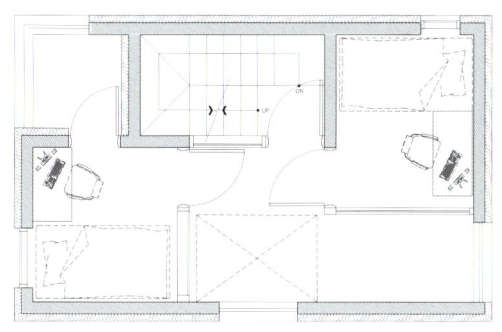
3RD FLOOR PLAN

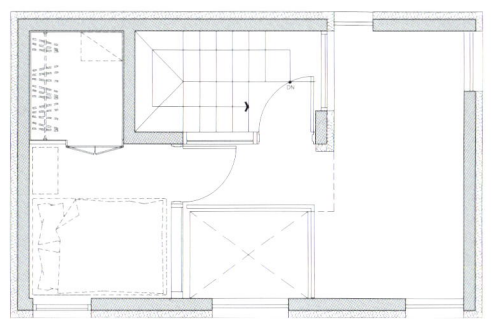
4TH FLOOR PLAN

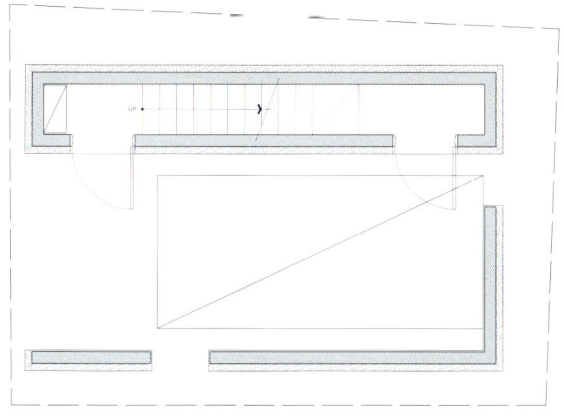
1ST FLOOR PLAN

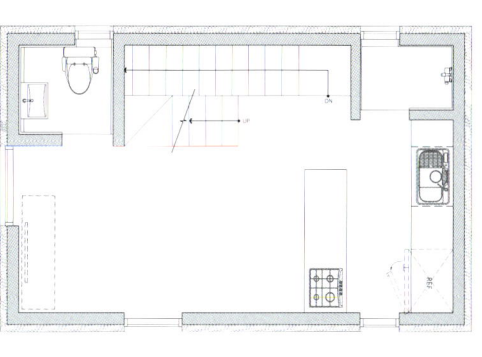
2ND FLOOR PLAN

31.61m²

Louver Top-light

Location
Nagoya city Aichi, Japan
Use
House
Site Area
43.45m^2
Built Area
31.61m^2(1F), 28.39m^2(2F), 30.85m^2(3F)
Total Floor Area
90.58m^2

Floor
3F
Structure
Steel
Exterior Finish
Autoclaved lightweight aerated concrete
Interior Finish
Oak Flooring,
Emulsion Paint, Lauan

Project Architect
Satoshi Kurosaki
Photographer
Masao Nishikawa

NEST

도심 속 새둥지 같은 아늑한 소형주택 네스트

APOLLO Architects & Associates

$25m^2 - 35m^2$

> 두 배로 높은 생활 공간 위의 상단 창문을 통해 햇빛이 충분히 들어오며, 하늘이 보이기 때문에 탁 트인 느낌을 준다. 발코니의 루버는 내부에서의 가시성을 제공하는 한편, 사생활 보호 수준 조절을 위해 외부로부터의 시선은 차단한다.

나고야역 근처의 상업 지역에 있는 이 소형 주택은 43m^{2}의 대지에 총 3층의 철골 구조로 지어졌으며, 건축 면적은 33m^2 미만이다. 갈색 루버와 회색 스프레이 페인트를 뿌린 외벽이 돋보이는 세련된 느낌의 건물 외관은 새 둥지를 연상시킨다.

지상 1층은 작은 차를 주차할 수 있는 필로티 스타일의 주차 공간, 진입로 끝에 있는 주택 출입문, 주 침실로 구성되며, 2층에는 아이들 방과 욕실만 있어 간결하다. 주 침실의 전체 공간은 공간이 비좁은 것을 고려하여 하나의 가구처럼 취급했다. 침대에는 수납형 매트리스가 있으며, 헤드보드에 장착된 간접 조명 장치는 이처럼 '비좁은' 공간에서만 가능한 아늑하고 친밀한 분위기를 연출한다.

2층에 있는 아이들 방은 의도적으로 계단 부분으로 이어지는 열린 배치로 만들어서 집 전체에 공간적 가변성을 더해줄 수 있는 다복적 공간으로 사용될 수 있게 했다. 이 주택은 매우 작게 구성되었기 때문에 단순한 기능성을 넘어서는 적합한 공간을 제공할 필요가 있다. 조종석 스타일의

주방은 장래에 어머니와 딸이 함께 요리를 즐길 수 있도록 지상 3층에 마련됐다. 소형 식탁과 연결되어 있는 이 주방은 집의 중심점 역할을 한다. 나왕 합판을 사용하여 빌트인 주방과 인테리어를 일관성 있게 마감 처리한 덕분에 작은 공간 내에서 통일감이 느껴지며, 세련된 디테일은 기분 좋은 느낌을 준다.

두 배로 높은 생활 공간 위의 상단 창문을 통해 햇빛이 충분히 들어오며, 하늘이 보이기 때문에 탁 트인 느낌을 준다. 발코니의 루버는 내부에서의 가시성을 제공하는 한편, 사생활 보호 수준 조절을 위해 외부로부터의 시선은 차단한다. 이는 노시 수택의 선형으로, 부지 조건을 최대한 활용하여 고객만의 독창성을 구현했다.

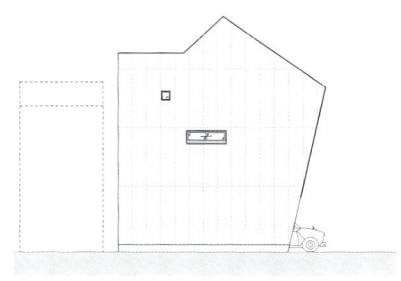
EAST ELEVATION

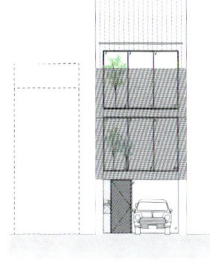
NORTH ELEVATION

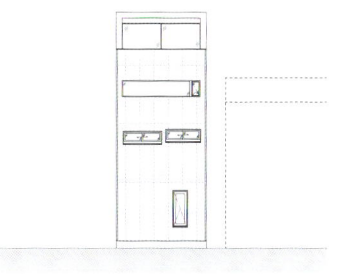
SOUTH ELEVATION

The top light above the double-height living space provides ample daylight, creating an open atmosphere with the view to the sky. Louver of the balcony allows visibility from inside, while the sight from the outside is blocked to control the privacy level.

This small house located within a commercial area near Nagoya station is built on a lot with 43m^2 in steel frame structure of three levels, while building area is less than just 33m^2. The chic building exterior, covered with brown-colored louver and exterior wall with gray-colored spray paint, looks as if like a nest of a bird.

The 1st floor level consists of a garage space in a piloti style for a small car, the entrance to the house at the end of the approach, and a main bedroom, while the 2nd floor level is compactly provided with a children's room and water section. The entire space of the main bedroom is treated as furniture to be conscious about the tightness of the space, thus a bed with storable mattress was provided along with a headboard equipped with indirect lighting system, creating a cozy and intimate atmosphere only available within such "tight" space.

The children's room on the 2nd floor level is intentionally kept as an open layout continuous to the staircase area, allowing to be used as a multi-purpose space for adding spatial flexibility to the entire house. Because of this house in such small configuration, it is necessary to provide a niche space beyond pure functionality. A cockpit-style kitchen is provided on the 3rd floor level for mother and daughters to enjoy cooking together in the future, while this kitchen is connected to a compact dining table to make the space as a central focus of the house. The finish of the built-in kitchen and the interior is consistent by the use of lauan plywood, creating a sense of unity within the small space and pleasant feeling from its fine details.

The top light above the double-height living space provides ample daylight, creating an open atmosphere with the view to the sky. Louver of the balcony allows visibility from inside, while the sight from the outside is blocked to control the privacy level. This is the prototype of an urban residence, realizing the originality of the client by making the best use of the site conditions.

$25m^2 - 35m^2$

25m² - 35m²

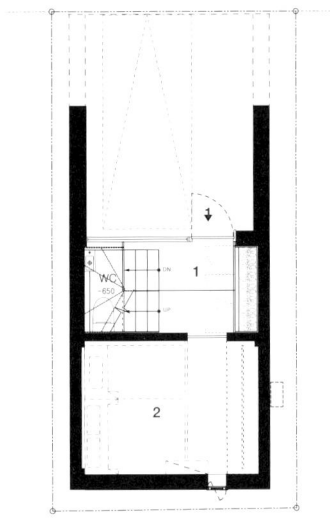
1ST FLOOR PLAN

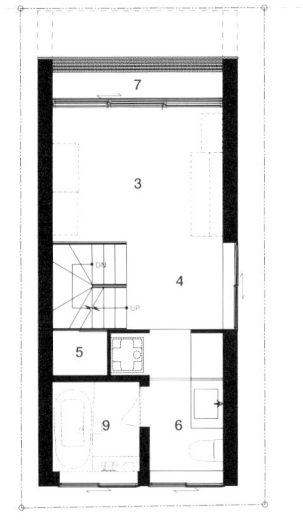
2ND FLOOR PLAN

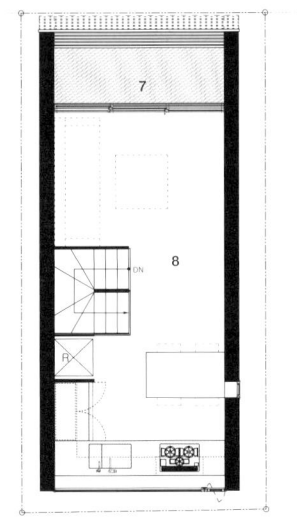
3RD FLOOR PLAN

SECTION

1 ENTRANCE
2 BED ROOM
3 CHILDREN ROOM
4 STUDY CORNER
5 STORE ROOM
6 WASH ROOM
7 BALCONY
8 LDK
9 BATH ROOM
10 ENT. HALL
11 WC
12 ROOF

NEST

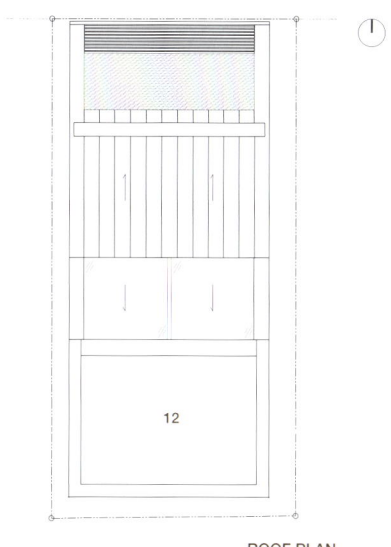

ROOF PLAN

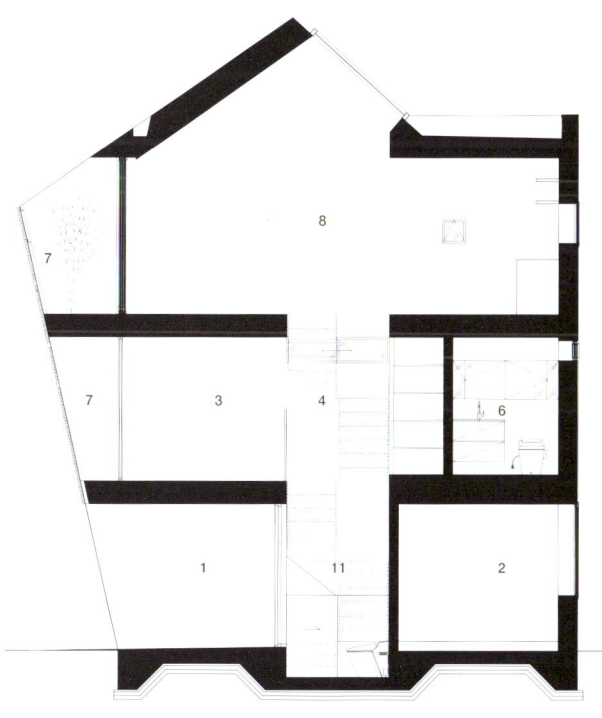

SECTION

28.77m²

Skipfloor
Various window

Location
Seoul, South Korea
Use
House
Site Area
56.2m²
Built Area
28.77m²
Total Floor Area
133.92m²

Floor
B1 - 5F
Structure
Reinforced concrete
Exterior Finish
STO

Project Architect
Cho Hanjun
Photographer
Ryoo Inkeun

[Crevice] 1740

거대한 도시 속 작은땅, 새로운 기회
틈 1740

ThePlus Architects (건축사사무소 더함)

25m² - 35m²

[Crevice] 1740

> 빛을 받아들이는 창의 형태나 채광방법, 내부 공간의 틈을 통해 층간의 단절이 아닌 연속적으로 이어지는 공간감, 그 연속성을 통해 정말 작은 내부공간이 최대한 답답하게 보이지 않게 하려고 하였다. 내가 서있는 장소에서 위 아래 플로어의 동선을 인지할 수 있으며 시각적인 연계를 유지하였다.

"거대한 도시안에서 작은 땅은 도시안의 '틈' 작은땅은 새로운 기회이다. 작은 틈으로 새어 들어오는 한줄기 '빛'같은 것".

나는 이 프로젝트의 이름을 '틈'(CREVICE)으로 부르자고 했다. 거대한 도시 속 작은 땅이 그 '틈'이었고 '틈'이라는 말은 때에 따라 '어떤 행동을 할만한 기회'라는 말로도 쓰이며 '시간적인 여유' 로도 쓰이고 물리적으로는 어떤 물체에 '벌어져 사이가 난 자리'이며 한줄기 가느다란 밝은 빛이 들어올 수 있는 그 '틈'을 의미하는 말로 부르고 싶었다. 그렇게 화두를 던지고 난 뒤 일관되게 그 '틈'을 계획의 콘셉트로 사용하였다. 빛을 받아들이는 창의 형태나 채광방법, 내부 공간의 틈을 통해 층간의 단절이 아닌 연속적으로 이어지는 공간감, 그 연속성을 통해 정말 작은 내부공간이 최대한 답답하게 보이지 않게 하려고 하였다. 내가 서있는 장소에서 위 아래 플로어의 동선을 인지할 수 있으며 시각적인 연계를 유지하고 싶었다.

서울에서도 대표적인 저층주거 밀집지역인 서대문구 천연동에 위치한 56.2㎡ 규모의 특이한 모양의 땅이다. 주어진 프로그램은 지하층에 작업실을 포함한 단독주택이다. 젊은 부부에게는 초등학생 딸이 있고 아이를 돌봐줄 도우미의 공간도 필요한 주택이다.

땅은 작지만 지어져야 할 주택의 쓰임은 결코 작지 않다. 젊은 부부는 직업적인 특성상 그리고 아이의 양육상 서울이라는 도시를 떠나 집을 지을 수 있는 여건이 안되있다. 이들이 결정한 것은 도시의 작은 자투리 땅에 집을 지어 주거를 해결함과 동시에 사회활동에도 제약을 받지 않으며 아이의 양육에도 제약을 받지 않는 그런 집을 짓기로 결정하였다. 작은 땅은 대부분 도로가 좁기 때문에 도로 건너편에는 다세대나 다가구 같은 집들이 마주하고 있다. 이 경우 정면의 창들은 서로에게 프라이버시를 침해할 수 있는 민망한 일들이 종종 벌어진다. 그래서 정남쪽 모서리의 틈으로 난 창으로 채광이 가능하게 하고 그 외 필요한 창들은 가급적 주변건물로부터 프라이버시를 확보할 수 있는 위치에 계획하였다. 건물의 진입도 대지의 형상을 활용하여 골목길 들어가듯 우회하여 진입한다. 주택 내부에 들어와서도 동선이나 각 공간의 효율성을 고려한 스킵플로어로 바닥들의 벌어진 틈을 통해 공간의 연속성이 반영 되게 하였다.

$25m^2 - 35m^2$

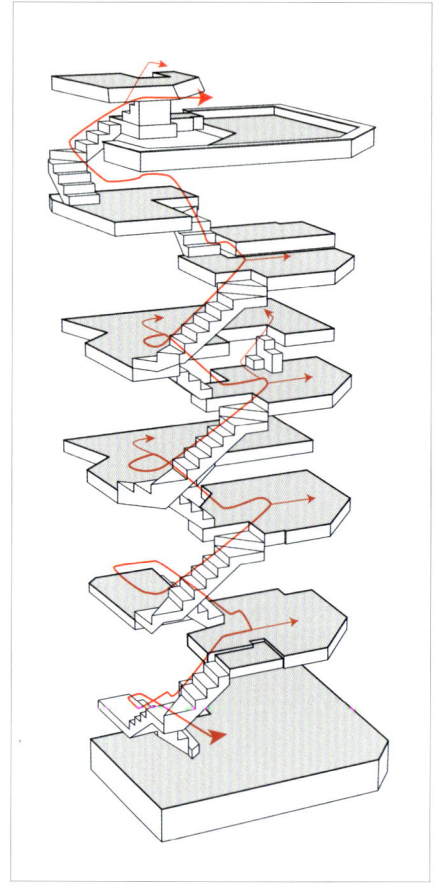

**With the shape of a window that receives light, the lighting method, and the sense of space that continues through the crevice of internal space without severance between floors, and that continuity, I tried to make the small internal space not look narrow.
I wanted to make a space where people could recognize the circulation up and downstairs from the point where he or she stood. I also wanted to maintain visual continuity.**

"A small piece of land in a huge city is a crevice. A small piece of land is a new opportunity. It is a ray of light that filters through a small crevice."
I wanted to call this project 'crevice.' A small piece of land in a huge city is a 'crevice,' and the word is also used to mean 'an opportunity to do some actions' and 'spare time.' I wanted to use it as a word that means 'a gap generated when something splits in two' or a chasm through which a ray of light is filtered. After throwing out a topic like that, I have used 'crevice' as the concept of the plan. With the shape of a window that receives light, the lighting method, and the sense of space that continues through the crevice of internal space without severance between floors, and that continuity, I

tried to make the small internal space not look narrow. I wanted to make a space where people could recognize the circulation up and downstairs from the point where he or she stood. I also wanted to maintain visual continuity.

This was a uniquely shaped 56.2m^2 lot in Cheonyeon-dong, Seodaemun-gu, a representative low-rise and high-density residential area in Seoul. The task given was to build a detached house that contained a studio on the basement floor. The young couple had a daughter who was an elementary school student, and the building also needed a space for a caregiver who cared for the child. The land area was small, but the use of the house to be built was not small. The young couple could not build a house outside Seoul because of the characteristics of their occupation and child rearing. So they decided to build a house on a small piece of land in the city, which gave them a residence and enabled uninhibited social activities and child rearing.

A small piece of land usually faces multi-household houses on the opposite side across a road that is narrow. In this case, people can see embarrassing things through the windows on the front of the building, which can infringe on privacy. So, I made a crevice of the corner on the south side for lighting and located other windows in a place where privacy from neighboring buildings could be secured. To enter the building, there is a detour like entering an alley using the shape of the land. Spatial continuity was reflected even inside the house through the crevice of grounds using a skip floor which considered the circulation or efficiency of each space.

$25m^2 - 35m^2$

1 WORKROOM
2 ENTRANCE
3 LIVING ROOM
4 BREAK FLOOR
5 PARKING LOT
6 LIVING ROOM
7 KITCHEN

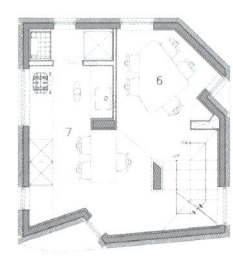

2ND FLOOR PLAN

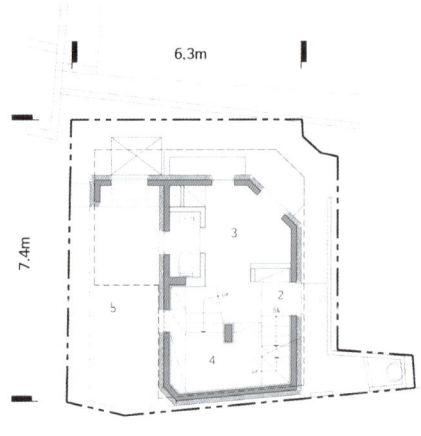

1ST FLOOR PLAN

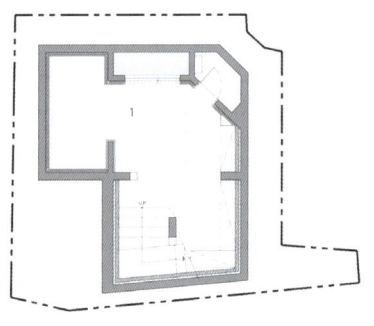

B1 FLOOR PLAN

[Crevice] 1740

1 ROOF	1 POWDER ROOM
2 STORAGE	2 ROOM
3 DRESS ROOM	3 WASH ROOM
4 KIDS ROOM	4 KITCHEN
5 MASTER ROOM	5 LIVING ROOM
6 KITCHEN	6 TOILET
7 PARKING LOT	7 WORK ROOM
8 WORK ROOM	
9 UTILITY	

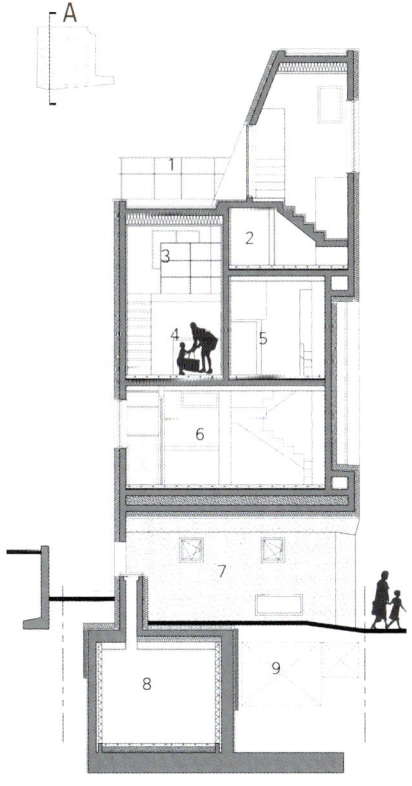

SECTION A

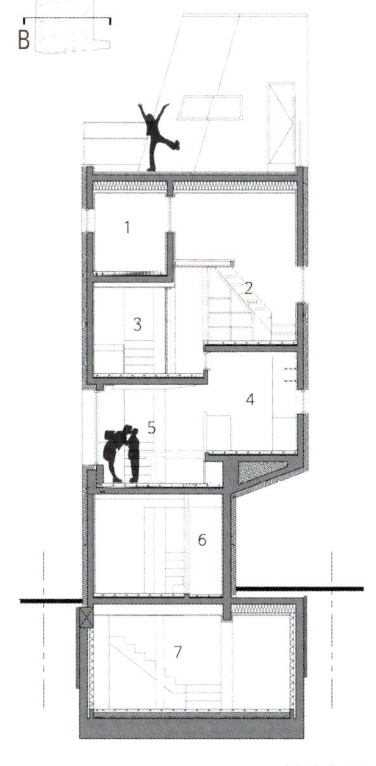

SECTION B

177

25m² - 35m²

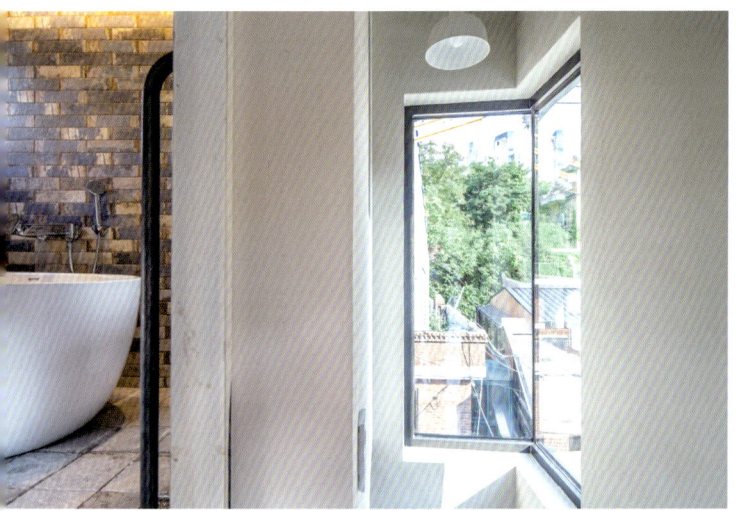

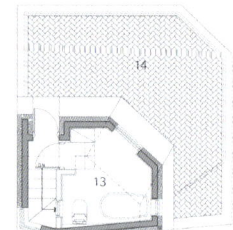

ROOF FLOOR PLAN

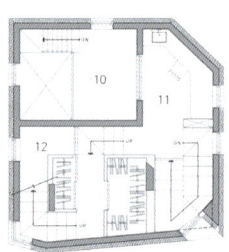

4TH FLOOR PLAN

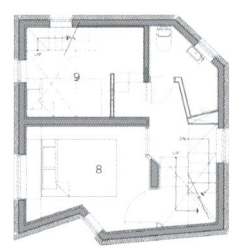

3RD FLOOR PLAN

8 MASTER BED ROOM
9 BED ROOM
10 LOFT
11 POWDER ROOM
12 STORAGE
13 BATH ROOM
14 ROOF

The plus Architect, Cho Hanjun

interview

효율적인 수직 공간구성을 위해서는 필요한 공간의 빈도와 중요도를 고려한 단면상의 조닝이 필요하다.

집은 거주자의 삶이 그대로 투영이 되는 거울이다.

Q. 협소주택의 정의를 무엇이라 생각하는가?
A. 협소주택이라는 단어에서 오는 어감의 불편함이 있다. 대부분의 협소주택을 지으려는 사람들에게 이 단어는 선택적이 않다.
자신들은 작은 땅에 지어야 하는 상황인데 그 안에서 자신들의 삶을 담아야 하는데 단지 땅이 작기 때문에 협소라는 말을 강제적으로 강요 받는 느낌이다.
우리가 집을 지을 때 물리적인 크기와 형태를 가지고 집에 대한 성격과 대한 정의를 내렸단 말인가. 다만 이러한 집들이 지어지는 현상에는 이러한 집을 지으려는 연령층과 그들의 경제적인 여건 그리고 도심지에서 집을 지어야 하는 라이프 스타일들이 협소주택에 대한 관심도가 높아진 것은 분명하다. 그러기 때문에 협소주택은 대부분 도심지 내에 지어진다. 비교적 저렴한 토지를 구하려고 하면 상대적으로 지가가 낮은 저층 주거단지 지역 내 작은 땅이 많다. 도로도 좁다. 그래서 공사를 하는 기간 동안에는 많은 주의가 필요하다. 협소주택은 단어가 가지고 있는 의미가 아니라 하나의 현상이고 물리적인 환경에 대한 특성이다. '협소주택'이라는 단어에서 굳이 의미를 찾고 정의를 내릴 필요가 없다.

Q. 작은 규모의 주택은 수직성에서 오는 불편함을 해결해야 한다. 이에 대한 아이디어는 어떠한 것이며, 프로그램의 조닝은 어떻게 하였는가?
A. 땅이 작기 때문에 집을 구성하는 필요한 공간들이 수직적으로 구성이 될 수 밖에 없다. 모든 공간들을 이동하기 위해서는 계단을 통해서 이동해야만 한다. 이러한 수직적인 공간의 구성을 최대한 효율적으로 하기 위해서는 필요한 공간의 빈도와 중요도를 고려하여 단면상의 조닝을 하여야 한다.

부부에게 좀더 자유로운 시간을 허락할 수 있는 작업실의 경우는 지하층에 배치하여 현관을 들어서자마자 바로 내려갈 수 있도록 하고 간혹 작업과 관련한 손님이 방문을 하더라도 사적인 공간을 거치지 않아야 할 것을 염두에 두었다. 1층의 경우는 주차장이 차지하는 면적 때문에 다른 층에 비해 공간이 협소하다. 1층은 작지만 현관을 들어서 마주하는 작은 홀의 영역이면서 지하층 작업실에서도 같이 사용해야 하는 화장실 수납공간이 주용도이다. 외출하기 전 마지막 공간이기 때문에 가급적 외출에 필요한 외투나 물건들은 이곳에 보관되어 있도록 하였고 잠시 숨을 고를 수 있는 공간이 될 수 있도록 의도하였다.
2층은 주방과 거실을 두어 이 집에서 가장 많은 시간을 보낼 수 밖에 없는 공간들이 위치하였다. 3층은 잠을 자는 공간인 안방과 아이방 그리고 화장실을 두어 2층과의 짧은 동선을 고려하였고 4층도 3층에서 바로 필요한 드레스룸이나 파우더룸으로 꾸며졌다. 아이방은 2개층의 층고를 활용하여 아이만의 작은 로프트도 고려하였다. 마지막 옥상의 욕실은 이 집에서 가장 높은 곳에 위치하였기 때문에 사용빈도는 작지만 그래도 가장 중요할 수 있는 욕실을 설치하여 욕조 안에서 휴식을 취할 수 있는 공간을 구성하였다.

Q. 집을 짓는 의미에 대해서 건축가로서의 생각이 궁금하다.
A. 집을 지으려는 사람들에게 있어서 그 이유와 동기는 모두 제 각각이다. 큰 틀에서 보면 주거로서의 '집'이라는 단어는 같지만 그 안에서 자신들의 삶을 담는 그릇으로서의 집이기 때문에 단순한 하드웨어를 만드는 일이 아닌 것이다. 벽을 세우고 지붕을 만들고 담을 두르는 물리적인 행위가 집을 짓는 과정의 전부일수 없다. 삶을 담아야 하는 집은 많은 이야기 거리와 집을 짓는 사람만의 라이프 스타일이 녹아들기 때문에 스토리를 엮어가는 과정이다. 그 엮어진 스토리를 바탕으로 공간이 구성이 되고 집의 형태가 완성이 되기 때문에 집을 짓는 다는 것은 하나의 생명체(유기체)를 만드는 일이다.

일상주택, 전원주택, 세컨하우스, 도심형 협소주택, 쉐어하우스 등 이러한 이름들은 거주하는 사람들이 어떤 식으로 집을 사용하느냐에 따라 나뉘는 이름이기도 하지만 그 내용이 집안에 고스란히 녹아 들기도 한다. 더 구체적으로는 같은 용도의 집이라 하더라도 모두 제 각각이다. 집은 거주자의 삶이 그대로 투영이 되는 거울인 것이다.
"집을 짓는 모든 이들에게 있어서 그 집은 이 세상에서 단 하나의 집이다."

The plus Architect, Cho Hanjun

interview

Q. What do you think "small houses" are?
A. There is some discomfort of nuance with the words "small house." These words are not helpful to people who want to build a small house. They have to build a house on a small piece of land and they have to contain their life in the house. Just because the land is small, it seems they are forced to use the word "small."

When we build a house, do we define the house and its characteristics by the size and shape of the house? However, it's obvious that interest in small houses has increased due to the age groups that want this kind of house, their economic situations, and the lifestyle of the urban area. That is why small houses are usually built downtown. If you want to purchase relatively low-priced land, there are many such small pieces of land in low-rise and high-density residential areas. However, the roads are also narrow. So, a lot of caution should be paid during construction. Small houses should be interpreted as a phenomenon with the characteristics of the physical environment instead of the literal meaning. In other words, there is no need to find meaning in the words "small house" and to define them.

Q. For small houses, discomfort from verticality should be solved. What are your ideas about this, and how did you zone the program?
A. As the land area is small, the spaces that compose the house cannot help but be built vertically. To move between spaces, people should use a staircase. To make the composition of such a vertical space efficient, cross-sectional zoning is needed considering the frequency and importance of the space required.

In the case of the studio that allows more free time for the couple, it was placed on the basement level so that the couple could go down immediately as soon as they walked through the front door. Also, I had in mind the fact that guests related to their work should not go through the private space when they visit the house.

The first floor is smaller than other floors because of the area of the parking lot. The first floor, although small, is a hall that is faced when people enter the front door and is composed of a bathroom on the basement level and a storage space.

[Crevice] 1740

As it is the last space in the house before going out, it was designed to store coats or other objects that are needed for going out. It was intended to be a place where people could catch their breath.

The kitchen and living room were placed on the second floor. The spaces people spend most time in are located here. The third floor has the main room, child's room, and bathroom, and the circulation between the second and third floors was considered. Also, the fourth floor was designed as a dressing room or powder room considering the fact that people could use the fourth floor immediately from the third floor. For the child's room, a small loft was built utilizing the story height of the two floors. Lastly, for the bathroom on the rooftop, a bathtub was installed to enable people to take a rest in it even though the use would not be frequent because it is located in the highest place in the house.

Q. What does it mean to build a house for you as an architect?

A. The reason and motivation for building a house are different for the people who want it. Although the word "house" is the same when looked at from a broader view, as a house is a

container for peoples' lives, it is not making hardware simply to build a house. Physical actions such as building a wall, making a roof, and putting up a fence cannot be all of the process of building a house. A house that contains life is a process of weaving a story, because the lifestyle of the people who build a house and various other stories melt into the house. The space is composed based on the story woven, and the form of the house is completed. Therefore, building a house means creating a life (organism).

General house, country house, second house, small urban house, and share house… these are made depending on the dwellers' use of houses, but actually the contents of the users' dwelling melt into the houses. More specifically, houses are all different, even if their use is the same. A house is a mirror that reflects the dweller's life.

"For all people who build a house, that house is the only house in the world."

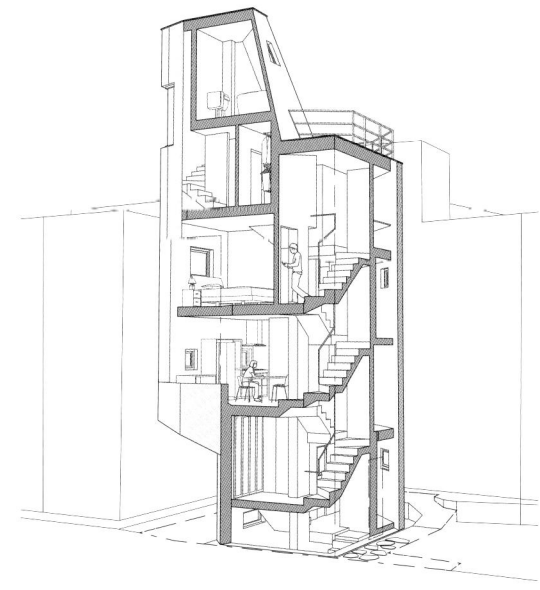

34.42m²

Floating Attic High ceiling

Location
Tokyo, Japan
Use
House
Site Area
87.08m²
Built Area
34.42m²
Total Floor Area
68.84m²

Floor
2F + attics
Exterior Finish
Mortar, wood bevel siding (cedar), metal
Interior Finish
Walnut, cedar, lauan plywood

Construction
Show Yo
Photographer
Masayoshi Ishii, Yuki Miyamoto

Small House with Floating Treehouse
떠있는 트리하우스를 담은 작은 집

Yuki Miyamoto Architect

Small house with floating treehouse

설계 콘셉트는 햇살, 바람, 녹지 등 기존의 천연자원을 최대한 활용해 건강한 주거 환경을 만들고 기술에 너무 의존하지 않는 디자인이었다. 덥고 습한 도쿄의 기후에서 자연 환기를 하려면 높은 천장이 더 적절했다. 두 개의 트리하우스 사이에는 창문을 배치했고 이 창을 통해 들어온 자연 채광이 거실을 포근하게 채워 준다.

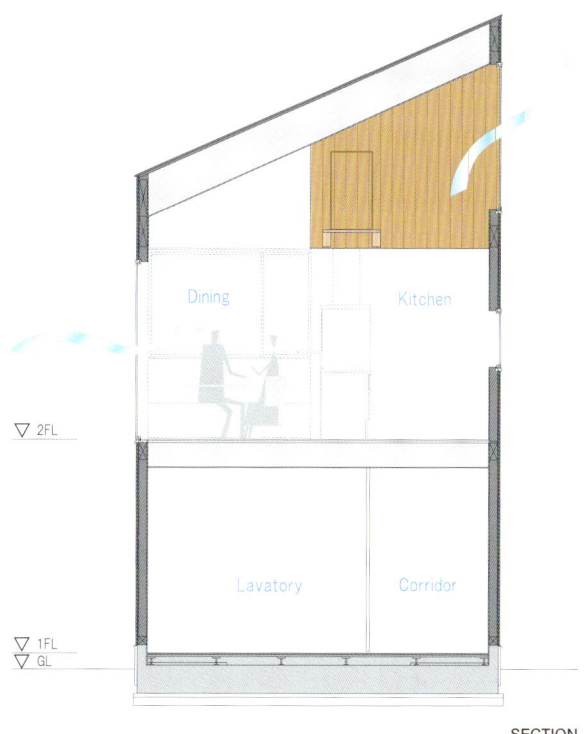

SECTION

이 주택은 도쿄 중심가의 경계에 자리한 조용한 수변가에 위치해 있다. 너비가 6.5m, 길이가 13m인 대지 위에 세워졌고 수변에는 아름드리 벚꽃나무를 포함해 수풀이 녹지가 풍부하다.
집의 소유주는 숲으로 캠핑을 하러 가는 등 야외 활동을 좋아한다. 시내에 세워진 집이지만 야외활동을 즐기는 의뢰인의 생활 방식을 반영하고 계절과 날씨의 변화를 가까이 느낄 수 있도록 설계해야 했다.
요구사항에 응하기 위해 다락에는 트리하우스처럼 만든 두 개의 방을 만들어 거실 위로 떠 있는 듯 보이도록 배치했다. 설계 콘셉트는 햇살, 바람, 녹지 등 기존의 천연자원을 최대한 활용해 건강한 주거 환경을 만들고 기술에 너무 의존하지 않는 디자인이었다. 덥고 습한 도쿄의 기후에서 자연 환기를 하려면 높은 천장이 더 적절했다. 두 개의 트리하우스 사이에는 창문을 배치했고 이 창을 통해 들어온 자연 채광이 거실을 포근하게 채워 준다.
평면은 강아지의 다리처럼 굽은 형상으로 설계했는데 이는 커다란 벚꽃나무를 바라보기 위해서였다. 이를 통해 우거진 녹음을 감상하면서 서쪽에서 불어오는 시원한 바람도 들일 수 있다. 2층은 35㎡로 면적이 좁은 편이지만 강아지 다리 형태가 공간에 시각적인 넓이와 깊이를 한층 더해 직사각형 평면보다 넓어 보이게 한다. 높은 천장과 공중 다리가 있는 구조 또한 공간을 수직으로 넓혀준다.
트리하우스의 존재로 인해 아이들에게는 재미있는 놀이터가 생겼을 뿐만 아니라 자연스럽게 따뜻한 분위기도 형성됐다. 동시에 야외 공간과 실내 공간의 경계가 흐려지고 실내에서도 야외와 같은 풍경을 느낄 수 있다.

Small house with floating treehouse

The concept of the design is to maximize the use of existing natural resources like, Sunlight, Wind and Greenery, to create a healthy residential environment and not depending too much on technology.

To make use of the natural ventilation, high ceiling is more appropriate for the hot and humid climate of Tokyo. And the windows set between 2 treehouse fill the living room with soft natural lighting.

SITE PLAN

The house is located in the quiet residential area in the boundary of Central Tokyo. It was built on a lot measuring 6.5m wide and 13m deep, surrounded by rich greenery including a big old cheery blossom tree.

The house owner likes outdoor activities, camping in a forest. Although this building was built in an urban area, the design of this house needs to meet their demand which can respond to the outdoor lifestyle and feel the changes of seasons and the weather intimately.

Therefore, the attic has 2 treehouse design rooms that seem floating above the living room. The concept of the design is to maximize the use of existing natural resources like, Sunlight, Wind and Greenery, to create a healthy residential environment and not depending too much on technology. To make use of the natural ventilation, high ceiling is more appropriate for the hot and humid climate of Tokyo. And the windows set between 2 treehouse fill the living room with soft natural lighting.

The typical plan was bent like a dog-leg, which was designed to face the big old cherry blossom and enfold the rich greenery and intake cool wind from the west. In addition, although the area on the 2nd floor is quite small, 35m², the dog-leg shape can give more expanse and depth to the space visually than a rectangular plan. The high ceiling and the existence of the floating bridge can also add a vertical expanse.

The existence of treehouse provides not only fun playgrounds for kids but also a natural warm atmosphere. At the same time, it can make the line between outdoor and indoor blurred and it creates an outdoor-like scene, while indoor.

25m² - 35m²

Small house with floating treehouse

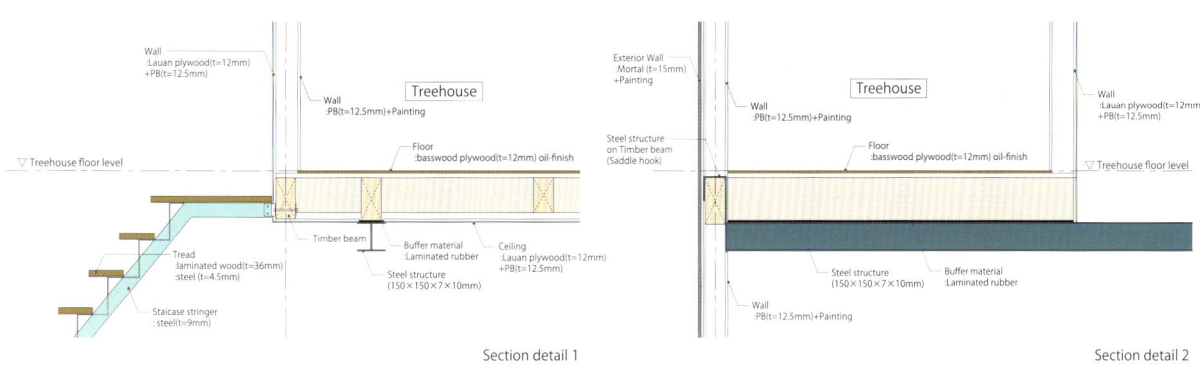

Section detail 1

Section detail 2

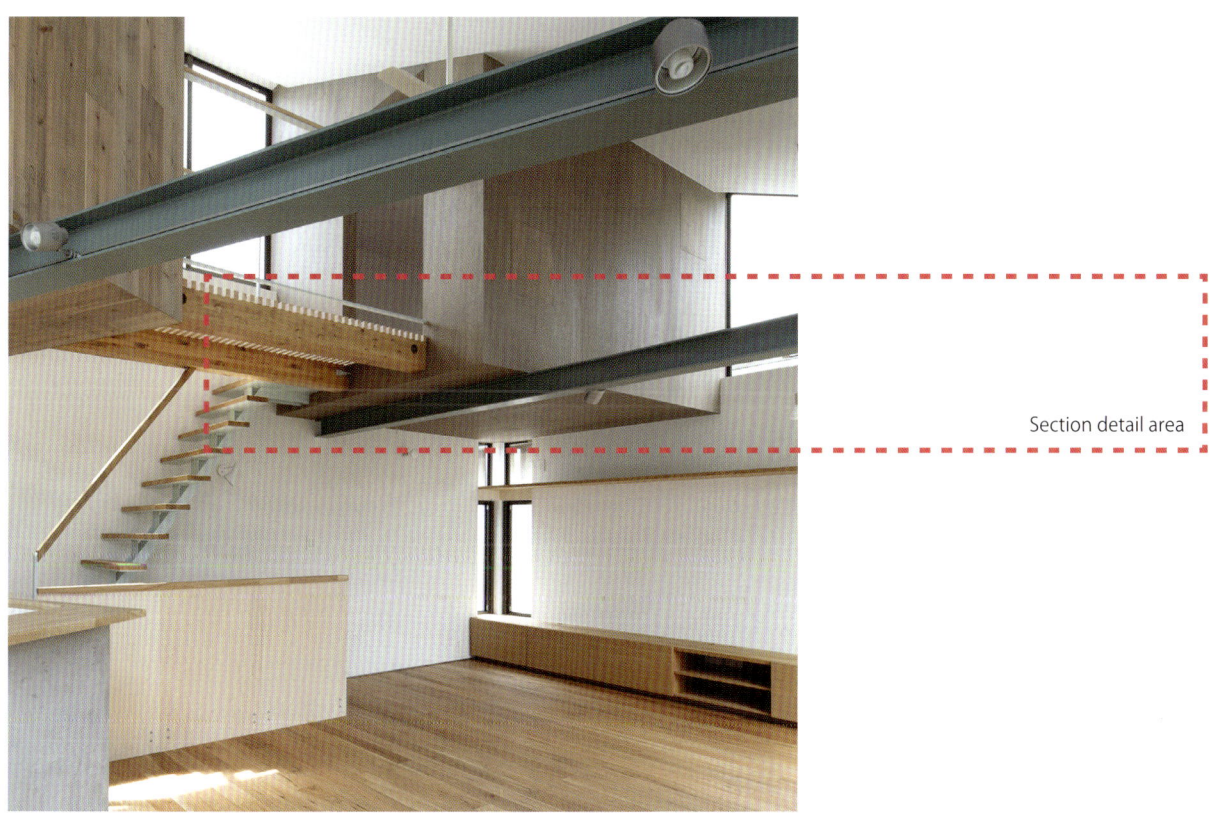

Section detail area

25m² - 35m²

1ST FLOOR PLAN

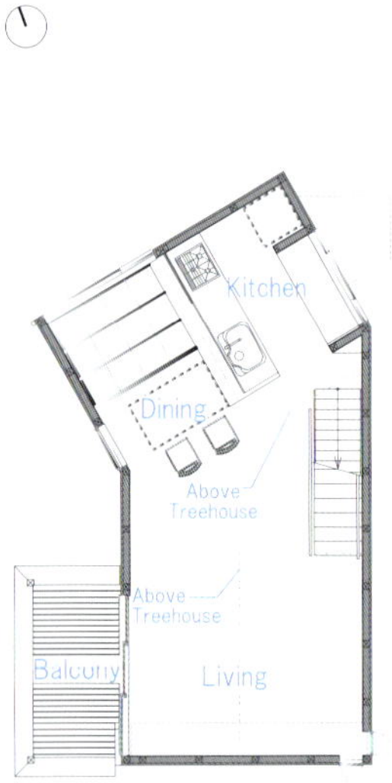

2ND FLOOR PLAN

Small house with floating treehouse

32.77m²

Terrace Unit Various window

Location
Tokyo, Japan
Use
House + hair salon
Site Area
181.92m²
Built Area
32.77m²
Total Floor Area
116.18m²

Floor
4F
Exterior Finish
Wood siding
Interior Finish
Paint

Project Architect
Hiroshi Kuno
Construction
Hirakata Komuten
Photographer
Hiroshi Kuno + Associates

Sixteen Rooms
거주자의 개성을 담은 16개의 작은방

Hiroshi Kuno + Associates

$25m^2 - 35m^2$

Sixteen rooms

네 개씩 네 층으로 쌓아 올린 16개의 작은 방은 각자 자신만의 방식으로 계절과 시간에 따라 내부 환경과 안락함을 만들어낸다. 거주자는 이곳에서 생활하며 각 방의 특성과 이를 이용하는 방법을 직접 알아낸다.

땅과 하늘, 뜨고 지는 태양, 중력이 만들어낸 바람, 주변의 식물이 선사하는 초록빛, 이 모두가 모여 환경을 이룬다. 건축가는 이러한 요소에 의해 완성된 환경에 그대로 화답하는 상자 모양의 건축물을 세웠다. 그리고 그 과정에서, 설계자의 해석에 따라 거주자를 안내하고 이끄는 건물이 만들어지기 직전에 멈추고자 했다. 자연 경관, 우주, 그리고 우리 자신의 존재와 마찬가지로, 자연이란 해석된 무언가가 아니다. 자연은 해석 이전의 무언가이며, 해석의 대상이다. 건축물이 그러한 자연과 좀 더 비슷해진다면, 건축물에 의해 형성되는 세상 역시 변화할 것이다.

네 개씩 네 층으로 쌓아 올린 16개의 작은 방이 있는 이 작은 집은 높이 솟아 있도록 설계되었다. 창의 크기와 높이가 각 방에 독특한 개성을 부여한다. 이 개성은 창의 방향과 높이, 주변 나무로부터 자연스럽게 얻어지는 것이다. 16개의 작은 방은 각자 자신만의 방식으로 계절과 시간에 따라 내부 환경과 안락함을 만들어낸다. 거주자는 이곳에서 생활하며 각 방의 특성과 이를 이용하는 방법을 직접 알아낸다.

16개의 방에는 커다란 창이 있어 탁 트여 있으며 나무 위의 집처럼 나무로 둘러싸인 방, 나뭇잎 사이로 흩어져 들어오는 빛으로 가득한 방, 서늘한 밤공기를 빨아들이고 밖으로 내뱉는 굴뚝 같은 방, 풍부한 빛을 받아 이를 다른 방과 공유하는 방, 그렇게 공유된 부드러운 빛으로 채워진 방, 지붕이 없는 방, 그리고 그 너머 가장 먼 방 등이 포함된다.

방이 작다면 태양과 중력에서 얻은 저밀도 에너지만으로도 환경이 크게 바뀔 수 있다. 모든 방이 항상 안락한 것은 아니지만, 언제나 어딘가에는 안락한 방이 존재한다. 모든 방을 안락하게 만들기 위해 많은 양의 에너지를 사용하는 것이 아니라, 사람이 움직여 안락한 공간을 찾는다. 그리고 이러한 자연스러운 행동이 집에 흥미로움을 선사한다.

종소리는 일단 발생하면 온전히 자율적으로 세상을 향해 나아간다. 그 소리에서 무엇을 느끼고 이를 어떻게 해석할지는 듣는 사람의 몫이다. 건축물이 해석 이전에 존재하는 무언가가 될 수 있다면 종소리와 같은 자율성과 추상성을 가지게 되며, 이로써 우리가 사는 세상은 좀 더 자유로워질 것이다.

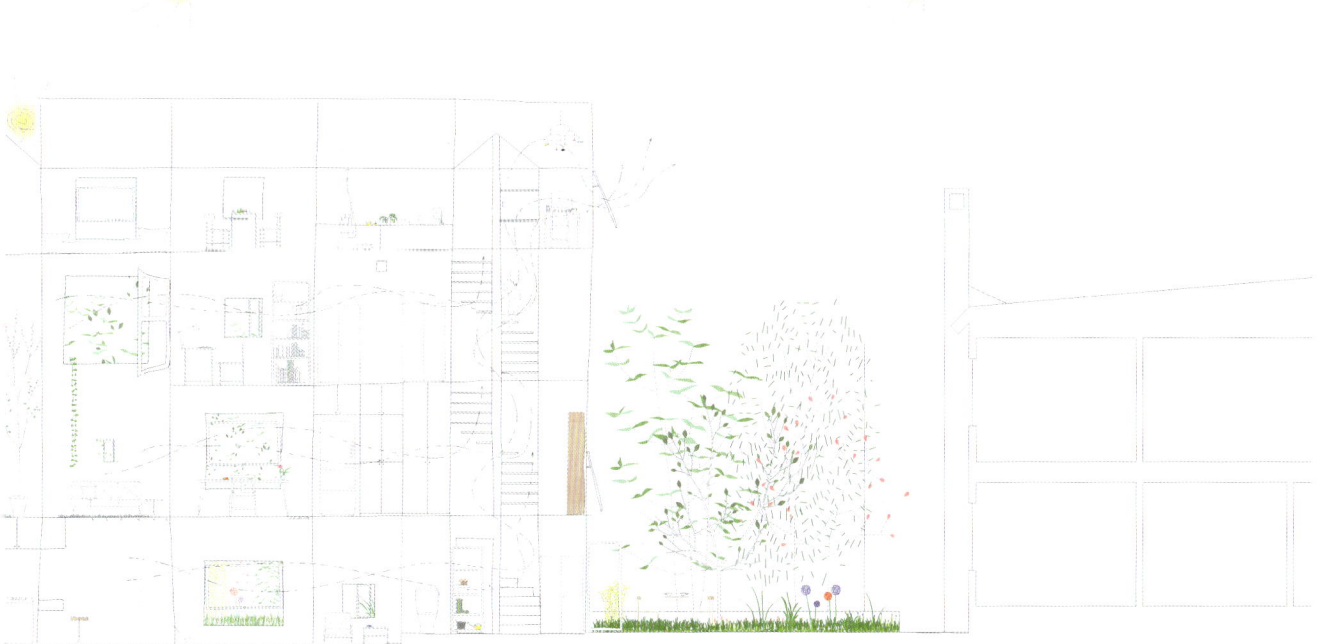

25m² - 35m²

Sixteen rooms

This small house of sixteen small rooms stacked up, four at a time in four layers, produces an interior environment and comfort in accord with the seasons and time. In living there, the residents discover the rooms and how to use them.

Ground, sky, the orbiting sun, and breezes caused by gravity, with the added greenery of peripheral plantings, engenders an environment. Here, I produced architecture in the form of a box that responds simply to the environment created by these elements. In this way, I felt it would be possible to stop one step before obtaining a building that orchestrates and instructs the residents according to the designer's interpretation. Like the natural landscape, like the universe, and like our own existence, nature is not something interpreted; it is something prior to interpretation, something that becomes an object of interpretation. When architecture is more like nature, as such, the world generated by architecture will also change.

This small house of sixteen small rooms stacked up, four at a time in four layers, is designed to stand high. The sizes and heights of its windows draw out the unique character that each room naturally obtains from its direction, height, and peripheral trees. The sixteen small rooms each in their own way produce an interior environment and comfort in accord with the seasons and time. In living there, the residents discover the rooms and how to use them. The sixteen rooms include a room opened with a large window and fitted like a treehouse among trees, a room filled with light diffused through trees, a room like a chimney that sucks in cool night air and conveys it outside, a room that takes in abundant light and shares it with other rooms, a room entirely filled with such soft light, a roofless room, and beyond it, the farthest room.

If a room is small, then even low-density energy obtained from the sun and gravity can change its environment greatly. Not all the rooms are always comfortable, but always, there is a comfortable room somewhere. Instead of using large amounts of energy to make all the rooms comfortable, the user moves about and finds a comfortable place, and this natural behavior gives enjoyment to dwelling.

Once emitted, the sound of bell goes out into the world on its own, entirely autonomous. It is we who choose what to feel in that sound and how to interpret it. If architecture can be something that exists prior to interpretation, having autonomy and abstractness like the sound of a bell, it might make our world a little freer.

25m² - 35m²

Sixteen rooms

SECTION DETAIL

Sixteen rooms

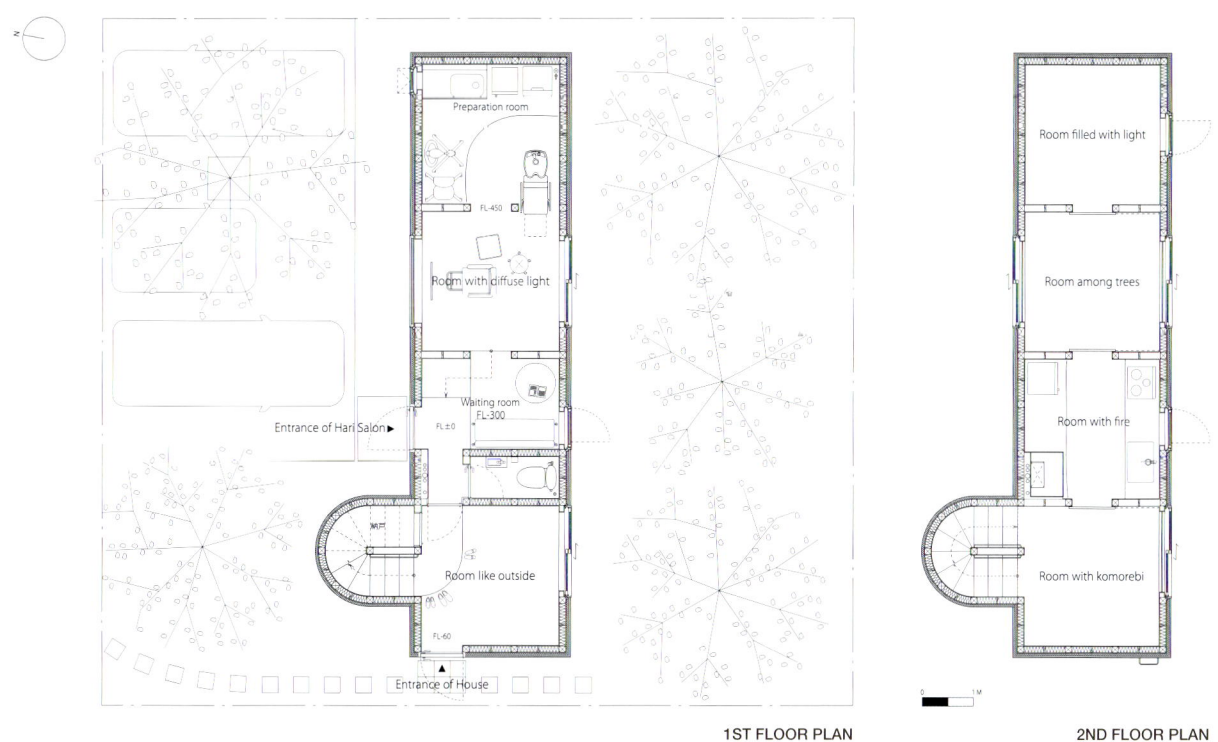

1ST FLOOR PLAN 2ND FLOOR PLAN

25m² - 35m²

Sixteen rooms

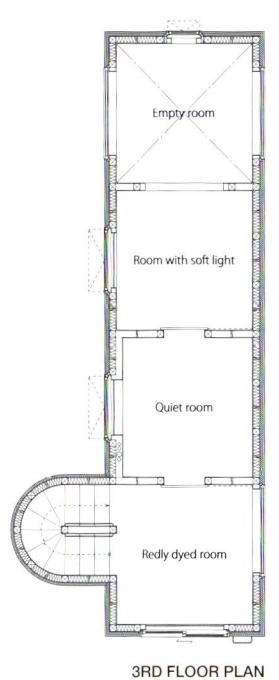

3RD FLOOR PLAN

4TH FLOOR PLAN

30m²

Prefabricated Toplight

Location
Beijing, China
Use
House
Site Area
30m²
Built Area
30m²
Total Floor Area
30m² (Plugin part_ 15.5m²)

Project Architect
He Zhe, James Shen, Zang Feng
Photographer
People's Architecture Office, Gao Tianxia

Mrs. Fan's Plugin House

천창을 통해 빛으로 가득 채운 판씨네 플러그인 하우스

People's Architecture Office

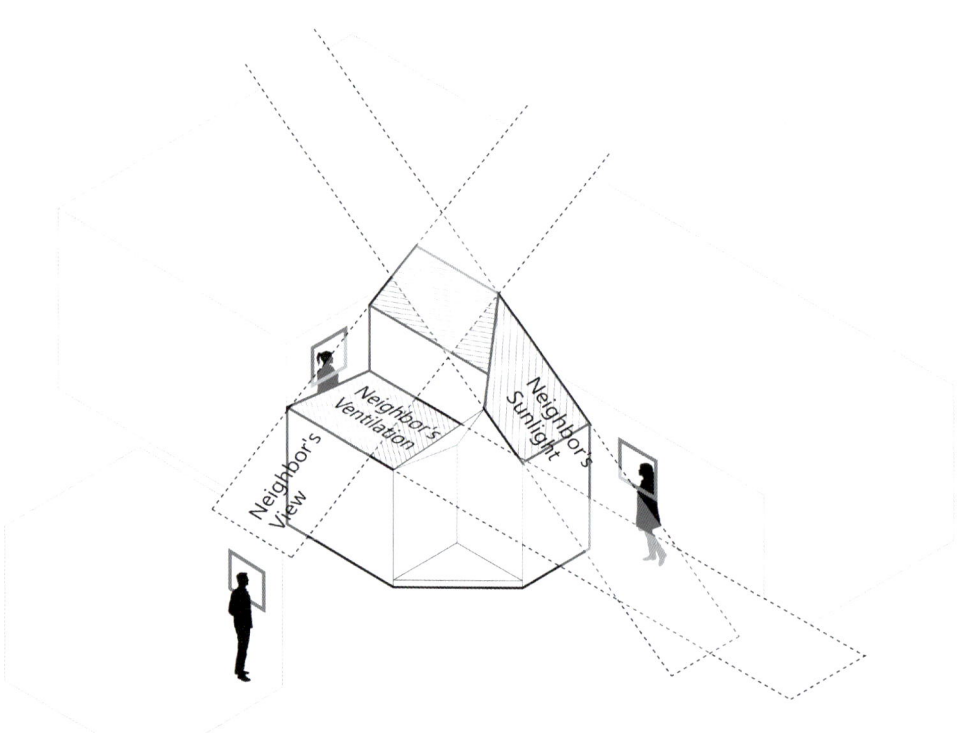

FORM OF PLUGIN HOUSE DEFINED BY NEIGHBORS

플러그인 하우스는 판 씨를 위해 맞춤형으로 설계되었다. 거실 부분의 천장을 높여 공간 높이를 두 배로 늘리고 양쪽에 천창을 냈다. 햇빛이 위에서 바닥까지 전달되어 이전엔 어두웠던 공간을 빛으로 채운다. 이 조립식 모듈은 단열재와 내외부 마감재가 몰드 패널 하나에 모두 통합되어 있는 형태이다.

클라이언트인 판 씨는 전통적 중국 가정 출신이다. 판 씨 부부 같은 신혼부부는 자동차를 사고 교외의 새집으로 이사하여 인생의 새로운 단계를 시작할 것으로 기대된다. 그러나 베이징의 천문학적인 부동산 가격을 생각하면, 재정적 독립을 원하는 30대 초반이 자가주택을 마련하는 것은 거의 불가능한 일이다.

판 씨는 베이징 옛 시가지 중심의 창춘지(Changchun Jie) 골목에서 나고 자랐다. 판 씨가 고등학교에 다닐 무렵 가족이 교외로 이주했으며, 고향 동네는 낡은 기반시설과 인구 과밀로 점점 슬럼화되었다. 하지만 판 씨는 교외의 고층주택에 결코 익숙해지지 않았으며, 어린 시절에 경험한 가까운 이웃 간의 친밀한 분위기를 그리워했다.

일반적인 아파트 구매가의 30분의 1 수준인 플러그인 하우스(Plugin House) 가격 덕분에, 자신이 자라난 동네로 돌아가고 싶다는 판 씨의 바람을 실현할 수 있게 되었다. 플러그인 하우스의 생활 수준과 에너지 효율은 신축 아파트 이상이다. 통근시간도 하루 4시간에서 1시간으로 줄어들었다. 플러그인 하우스는 옛집의 일부를 교체하고 주방과 욕실 등 새로운 기능을 추가하여 완성되었다. 창춘지 골목에는 하수시설이 없기 때문에, 보통은 공중화장실을 이용할 수밖에 없다. 그러나 플러그인 하우스는 하수도 연결이 불필요한 자연발효식 화장실을 설치하여 좀 더 쾌적한 생활이 가능하다.

$25m^2 - 35m^2$

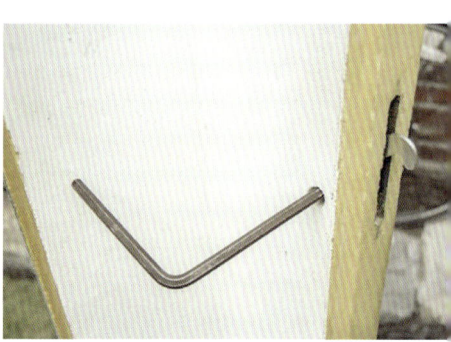

플러그인 하우스는 판 씨를 위해 맞춤형으로 설계되었다. 거실 부분의 천장을 높여 공간 높이를 두 배로 늘리고 양쪽에 천창을 냈다. 햇빛이 위에서 바닥까지 전달되어 이전엔 어두웠던 공간을 빛으로 채운다.

폐소공포증이 있는 판 씨를 위해 작은 욕실에도 천창을 냈으며, 프라이버시 보호를 위한 푸른색 차단막을 통해 반사된 햇빛이 들어온다. 흐린 날에도 욕실은 푸른 빛으로 덮인다. 옥상 데크는 북적이는 환경을 잠시 벗어나 숨을 돌리는 휴식 공간이자 전용 손님맞이 공간이 된다.

BEFORE

AFTER

피플즈아키텍처오피스(People's Architecture Office)만의 조립식 플러그인 패널 덕분에 플러그인 하우스의 건축비는 매우 저렴하다. 애초 코트야드하우스 플러그인(Courtyard House Plugin)의 "집 안의 집" 리노베이션을 위해 개발된 새로운 플러그인 하우스 시스템은 방수 기능을 갖추었으며 기존 구조 위에 설치할 수 있다. 이 조립식 모듈은 단열재와 내외부 마감재가 몰드 패널 하나에 모두 통합되어 있는 형태이다. 역시 패널의 일부인 고정장치를 이용해 플러그인 패널을 서로 연결하면 건설이 완료되며, 이는 비숙련자 두 명이 도구 하나를 이용해 하루 만에 끝낼 수 있을 정도로 간단한 작업이다. 배선과 배관 역시 복합 몰드 패널 안에 통합되어 있다.

플러그인 하우스의 건축 형태는 규제에 따른 제한이 아니라, 주변 이웃의 요구에 대한 타협을 바탕으로 정의된다. 플러그인 하우스의 구조는 어느 쪽으로든 이웃의 채광, 환기, 조망을 방해하지 않아야 한다. 시공이 완료된 후에도 새로운 요구가 발생할 수 있다. 플러그인 패널 소재는 건물 전체를 현장에서 절단할 수 있기 때문에 이러한 변화도 수용할 수 있다.

서로 교차하는 사회적 힘을 표현하는 플러그인 하우스는 지역적 조건에 맞춰 탄생한 새로운 도시형 토속건축 양식이다. 판 씨 같은 원래 주민이 베이징 옛 시가지로 돌아오는 것은 드문 일이다. 플러그인 하우스는 주어진 사회적 제약 내에서 저렴한 비용으로 생활 수준을 높여, 오래된 골목에 새로운 생명을 불어넣고자 한다.

1. Terrace
2. Entrance

SITE PLAN

The Plugin House is custom designed for Mrs. Fan. The living room ceiling extends upwards to provide a double height space with skylights on either side. Sunlight is channeled in from above to flood the previously dark interior with light. These prefabricated modules incorporate insulation, interior and exterior finish into one molded part.

Mrs. Fan is from a traditional Chinese family. Newlyweds like her are expected to purchase a car and move into a new house in the suburbs to start the next phase of their life. But for people in their early 30's who wish to be financially independent, the astronomical price of real estate in Beijing makes buying a house on their own nearly impossible.

Mrs. Fan was born and raised in the Changchun Jie Hutong neighborhood in the center of historic Beijing. By the time she was in high school her family had moved to the suburbs while her old neighborhood, with outdated infrastructure and overcrowding, continued to des cend into slum-like conditions. But Fan never got accustomed to suburban residential towers, preferring the intimacy of the close knit community she came from.

The affordability of the Plugin House, thirty times less than the cost of buying a typical apartment, made moving back to where Mrs. Fan grew up a practical reality. The living standard and energy efficiency of a Plugin equals or exceeds that of new apartment towers. And her daily commute to work is now reduced from four hours to one. The Plugin replaces part of the old house and adds new functions such as a kitchen and bathroom. The Changchun Jie neighborhood has no sewage system, so public toilets are usually the only option. But an off-the-grid composting toilet system integrated into the plugin makes Hutong life much more convenient.

BEFORE

AFTER

Mrs. Fan's plugin house

The Plugin House is custom designed for Mrs. Fan. The living room ceiling extends upwards to provide a double height space with skylights on either side. Sunlight is channeled in from above to flood the previously dark interior with light. To relieve Mrs. Fan of her claustrophobia the small bathroom also has a skylight but receives reflected sunlight from a blue privacy screen. Even on gloomy days the bathroom is covered in a blue tint. A roof deck gives her breathing room from the dense surroundings and private social space.

PAO's proprietary prefabricated Plugin Panels makes the Plugin House very affordable. Originally developed for the Courtyard House Plugin for "house in house" renovations, the new Plugin House System is waterproof and can be used outside of an existing structure. These prefabricated modules incorporate insulation, interior and exterior finish into one molded part. Plugin Panels attach to each other with an integrated lock making construction a task simple enough to be completed by a couple of unskilled people and one tool in one day. Wiring and plumbing are integrated into the molded composite panels.

The architectural form of the Plugin is defined not by limitations imposed from regulations but instead the negotiated demands from surrounding neighbors. On all sides of the Plugin the structure cannot block sun light, air circulation, and views of the people next door. Even as the structure was built, new demands came about. The Plugin Panel material makes accommodating these changes practical, chopping off entire sections of the building can be done on site.

As an expression of intersecting social forces the Plugin House is a new urban vernacular born from local conditions. For original residents like Mrs. Fan to move back to these historic parts of Beijing is rare. Through improving living standards for an affordable price within given social constraints the Plugin House attempts to breathe new life into old neighborhoods.

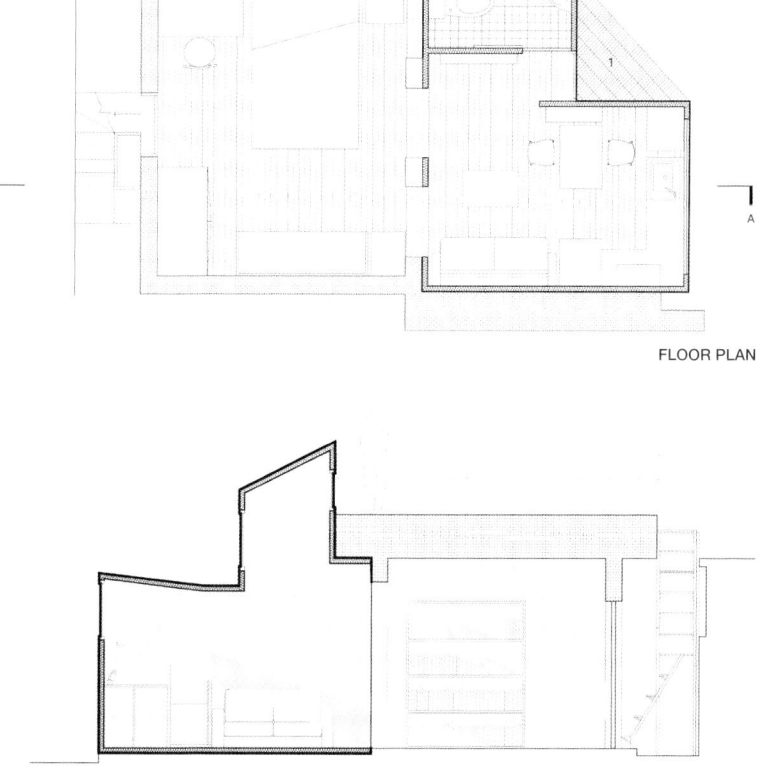

FLOOR PLAN

SECTION A

25m² - 35m²

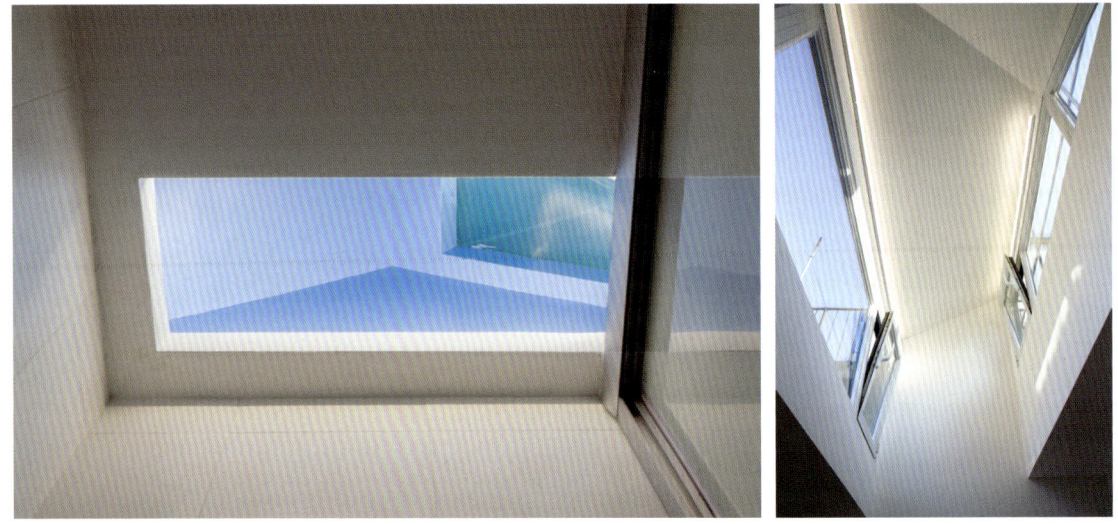

Mrs. Fan's plugin house

33.41m²

Toplight
Sliding door
Outdoor terrace

Location
Seoul, South Korea
Use
Housing
Site Area
60.45m²
Built Area
33.41m²
Total Floor Area
79.4m²

Floor
3F + loft
Structure
Reinforced concrete
Exterior Finish
STO

Project Architect
Park Changhyun
Construction
Cho Roksun
Photographer
Chin Hyosook

Seongsan-dong Mix-use Housing

골목과의 유동적인 관계를 만든 집
성산동 주택

a round Architects (에이라운드 건축)

25m² - 35m²

> 좁은 골목에서 나타나는 유연한 경계가 집과 집의 관계를 잘 보여준다. 우리는 좁은 골목에 접한 1층의 긴 경계를 담이나 벽으로 구성하지 않고, 투시가 가능하고 움직일 수 있는 슬라이딩 도어만으로 유동적인 경계를 만드는 것으로부터 시작하였다.

골목의 스케일

대지가 위치한 곳은 서울의 오래된 동네이다. 60년대 도시계획에 의해 주택단지로 필지가 구획되었지만 여전히 다양한 스케일이 공존하고 있고 시간이 누적되어 있는 동네이다. 가까운 거리에 8차선도로, 4차선도로, 5m도로, 그리고 마지막으로 세 집만 사용하는 2m의 막다른 도로까지 다양한 스케일의 길을 거쳐야 다다를 수 있는 작은 대지이다. 각 도로 폭의 사이는 단지 물리적인 사이만이 아니라 그 도로를 사용하는 사람들, 그리고 사람들끼리의 관계도 각각 다르다. 넓은 폭의 도로에서 지나치는 사람들은 같은 지역에 살지만 거의 알지 못한다. 그리고 서로를 알 수 있다는 의식 조차도 없다. 그렇지만 좁은 도로로 들어오면서 점점 사용자가 한정되고 그 한정된 사람들끼리의 관계는 서로 모른 척 하기란 쉽지 않다. 자주 만나게 되고 만나게 되면 인사하게 되기에 서로의 관계가 긴밀해진다. 게다가 일부 사람들만 사용할 수 있는 좁고 짧은 골목은 어느 정도 사유화 되어 그곳에서 보이는 풍경이나 사용성은 넓은 도로에서 보이는 모습과는 사뭇 달라진다. 자신의 집에서 키우던 화분이나 일부 개인의 물건들이 골목으로 나와 특이한 성격을 만들어 내기도 하고, 그 골목의 사용성도 그런 구성에 따라 달라지게 된다. 이런 모습들로 인해 집과 골목과의 영역 구분도 모호해 진다. 그런 관계는 서로의 관계를 느슨하게 만들어 내고 그것에 따라 섬세하게 그곳의 특징들이 드러난다.

골목과 집과의 경계

좁은 골목에서 나타나는 유연한 경계가 집과 집의 관계를 잘 보여준다. 우리는 좁은 골목에 접한 1층의 긴 경계를 담이나 벽으로 구성하지 않고 투시가 가능하고 움직일 수 있는 슬라이딩 도어만으로 유동적인 경계를 만드는 것으로부터 시작 하였다. 그리고 1층은 2층 보다 1.5m정도 물러나서 벽이 생기는데 이 벽의 2/3가 유리로 된 근린생활시설을 계획하였다. 이 근린생활시설은 지금 갤러리로 운영되고 있는데 작은 갤러리는 도로와의 관계를 슬라이딩 도어를 닫아 프라이버시를 확보할 수도 있고, 도어를 열어 도로가 확장되어 외부의 시선을 받아들일 수도 있다. 그리고 1층의 갤러리의 바닥은 골목의 바닥 높이보다 1m정도 낮춰 내부 천정 높이를 높게 계획함으로써 개방감을 최대화 시켰다. 그리고 골목의 한쪽에 면한 곳에 주택으로 들어갈 수 있는 현관 도어가 있는데 이곳 앞은 움직이는 슬라이딩 도어가 항상 막음으로써 벽의 느낌을 주어 프라이버시를 강조하였다.

주택의 구성

1층 현관에서 신을 벗고 계단으로 바로 올라간다. 이곳 계단도 천장이 높고 측벽은 막혀 있어 공간의 높이를 최대로 계획하였으며 천장과 만나는 벽에는 긴 고측창을 통해 서향의 오후의 빛을 내부로 받아들였다. 2층은 세 단계의 높이 차를 두고 계획하였다. 화장실과 부엌 조리공간은 낮고 두 계단 올라가서 거실과 다이닝이 함께 구성되어 있는데 이곳의 높이는 아래쪽 부엌에서 서 있는 높이와 거실에서 앉아 보이는 눈높이가 같다. 그리고 두 계단을 올라 거실의 가구 높이와 맞춰 계단을 구성하였는데, 이는 계단과 가구를 하나로 연결시키고자 하는 의도가 있었다. 다양한 높이 치이는 이곳에서 지랄 아이를 위한 제안이었고, 이는 이곳외 특징적인 공간의 요소가 될 것이다. 몇 개의 계단을 오르면 3층이 나오는데 이곳에는 서재, 화장실과 목욕실, 아이 방으로 구성되어 있다. 이 각각의 실들은 위 층까지 하늘을 향해 뚫여 있는 작은 중정을 중심으로 계획되었고 이 중정을 열면 환기와 채광이 적극적으로 기능하게 설계 되있다. 화장실과 목욕실 앞의 외부 데리스로 계획되이 빛을 받고 빨레를 널 수 있는 외부 공간을 마련하였다. 미지막으로 최상층에 부부침실을 두어 이곳에서 바로 나갈 수 있는 외부 테라스는 주택에서 맛 볼 수 있는 외부 공간이 있다.

작은집에서 채광과 한기에 대한 부분을 포함하여 다양한 공간이 구성은 이 집 만의 특징이다. 다양한 행위들이 일어나고 외부 주변 건물들과의 관계를 고려한 이곳은 이전에 경험한 훨씬 넓은 주거보다 다양하고 특별한 추억을 많이 만들어줄 것으로 기대한다.

Flexible boundaries found in narrow streets illustrate relationship between houses. We began the project with making flexible boundary; we designed the long boundary facing the narrow alley as sliding doors which allow visibility rather than fence or solid wall.

Scale of Village Alley

Construction site is located in old neighborhood in Seoul. Despite the plot-division into collective housing area by urban planning in 60's, variety of scales coexist and time has been gradually accumulated in the neighborhood. To reach the site, you have to go through various scales of roads; an eight-lane road, four-lane, five-lane, and lastly the closed-end 2m road which is only used by three households. Not only the physical scale, but also the people who use the road and the relationship between them differ from each other. People who pass by barely know about each other even though they live in same district, and furthermore don't make big effort to recognize their acquaintance. However when it gets to a scale of narrow street, it is not easy to just pass by someone you know. Mutual relationship gets stronger as people keep come across each other and exchange greetings. Narrow and short allies used by few neighbors become privatized, therefore street scenes and space use differ from wide roads. Flowerpots and personal belongings come out to the street and make unique sceneries. The use of the street changes accordingly. Due to these characteristics, distinction between house and street gets blurred due to these circumstances, and characteristics of the space come to the front.

Boundary of House and Street

Flexible boundaries found in narrow streets illustrate relationship between houses. We began the project with making flexible boundary; we designed the long boundary facing the narrow alley as sliding doors which allow visibility rather than fence or solid wall. The ground floor wall is set 1.5m back from the upper floor wall, and we composed the 2/3 of the wall with transparent glass. This space is currently used as gallery. This small gallery can secure its privacy by closing the sliding door, or draw people's attention by opening it. Openness in gallery space is maximized by raising the ceiling height through planning the floor level 1m lower than ground level. In case of the main door to residence which is facing the street, the sliding door conceals it in every condition in order to emphasize privacy.

Composition of Residence

You take off the shoe in the entrance and go right up to the stair. Ceiling is planned in maximum height since the both side of the wall are solid. Wide clerestory window located on the wall meeting the ceiling bring in western sun in the afternoon. Second floor consists of three floor levels. Bathroom and kitchen is located in lowest level and space for living room and dining is planned in next level which is two stair-step higher. The eye level in which you stand in the kitchen and sitting in the living room are the same. There is another two-step stair which has same height with furniture in living room with intent to connect the furniture and stair into one. Variety of level difference was proposed for the children who will grow up in the house, and this became the characteristic of the space. If you take few steps higher, you will be standing on third floor composed of study, bathroom, and kid's room. Each room is centered around the courtyard which is open to the sky. Natural ventilation and lighting are achieved through the courtyard. A terrace is located in front of bathroom, offering a exterior space for lighting and laundries. Lastly, a master bedroom is located on the top floor with a terrace accessible from the room, which is said to be a privilege of detached house.

Various spatial compositions, lighting, and ventilation are the characteristics of this house. Diverse activities taking place in the house and relationships between surrounding buildings are deeply considered. We wish this small house to make special and colorful memories in comparison with the large houses before.

1. STORE
2. KITCHEN
3. LIVING ROOM
4. ROOM

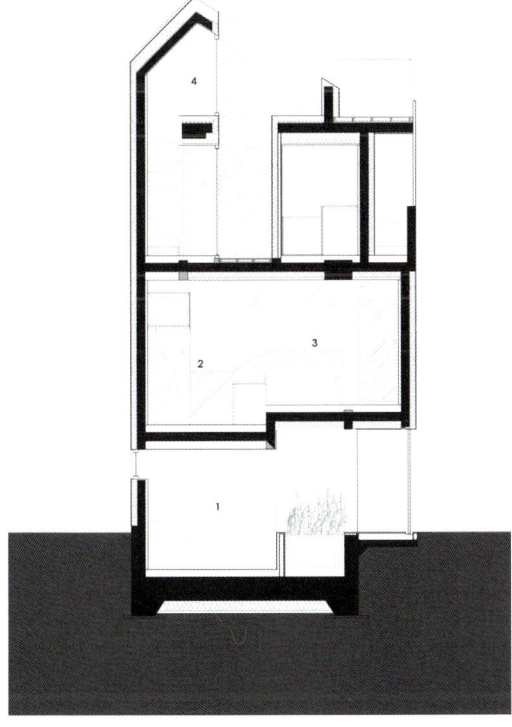

SECTION

$25m^2 - 35m^2$

Seongsan-dong mix-use housing

1 MAIN ENTRANCE
2 STORE
3 KITCHEN
4 LIVING
5 ROOM
6 BATHROOM

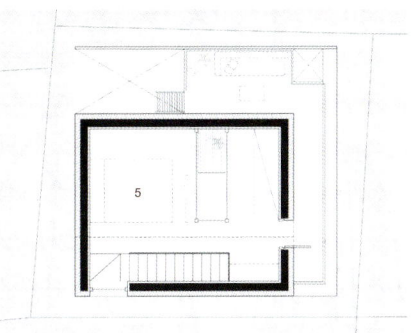

4TH FLOOR PLAN

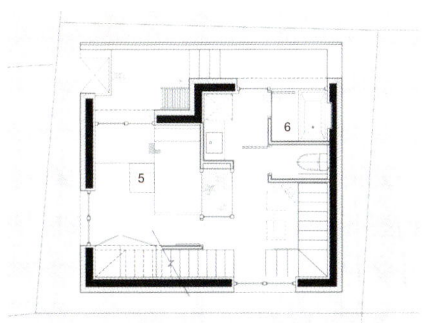

3RD FLOOR PLAN

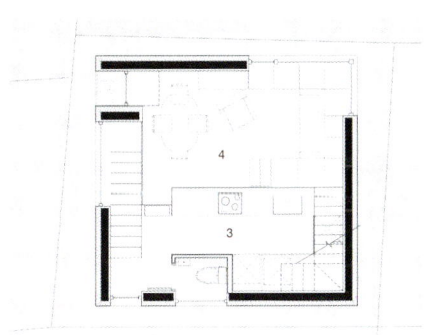

2ND FLOOR PLAN

1ST FLOOR PLAN

$25m^2 - 35m^2$

1 STORE
2 TOILET
3 ROOM

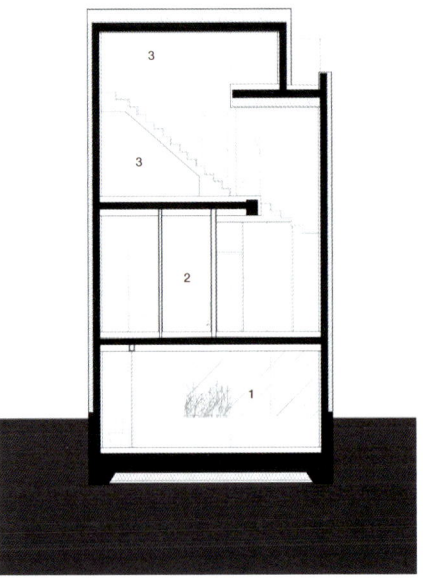

SECTION

Seongsan-dong mix-use housing

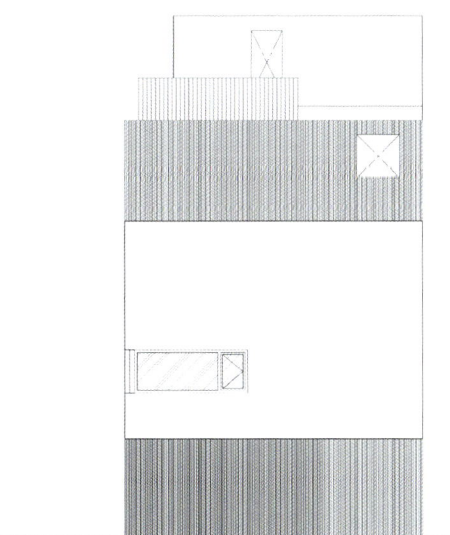
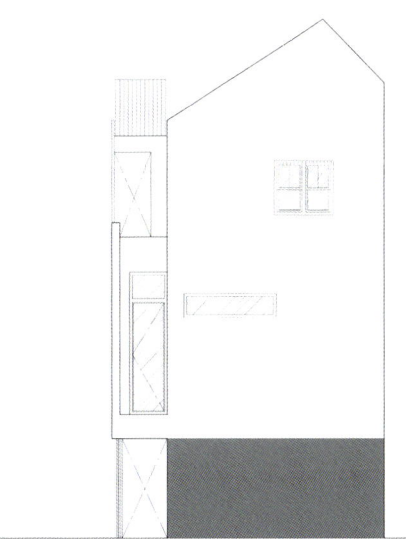

ELEVATION

a round architects, Park Changhyun

interview

대지의 면적이 작아서 그 속에서의 삶이 간결하게 되어야 한다기 보다, 번잡하고 펼쳐져 있는 삶을 정리하고 단순하게 사는 장소로써 작은 집이 협소주택이라고 생각한다.

Q. 본인이 생각하는 협소주택의 정의는 무엇인가?
A. 협소주택이 물리적으로는 작은 집일 가능성이 높지만 간결한 삶을 바탕으로 정의 내리고 싶다. 대지의 면적이 작아서 그 속에서 삶이 간결하게 되어야 한다기 보다, 번잡하고 펼쳐져 있는 삶을 정리하고 단순하게 사는 장소로써 작은집이라고 생각한다.

Q. 클라이언트의 요구사항은 무엇이었고, 그에 대한 건축적 대응은 어떤 것이었나?
A. 작은 땅의 협소주택이지만 수익이 생길 수 있는 상가를 함께 넣고 싶다는 요구가 있었다. 먼저 1층에 상가를 생각하게 되었지만 상가의 층고가 높아지면 주택의 최상층의 면적이 일조건 사선 제한에 의해 많이 줄어드는 부담이 있었다. 그래서 1층의 바닥 높이를 도로로부터 1m 정도 낮추어 1층의 면적은 작지만 높은 볼륨을 확보 할 수 있었고, 2층·3층의 주택 면적은 조금 더 넓게 계획 되었다.

Q. 협소한 면적 내에서 최대한의 공간 창출을 위한 아이디어는 무엇인가?
A. 먼저 1층 상가에서는 도로와 접하는 면에 열고 닫을 수 있는 목재 루버로 된 큰 도어를 만들어 도로와 1층 상가에서의 시선 확보를 하게 되었다. 주택에서는 2층 거실, 부엌, 식당을 한 공간에 계획하였지만 가구의 높이를 고려해 바닥의 단 차이를 두어 더욱 하나의 공간으로 사용할 수 있도록 계획한 부분이 공간의 쓰임새를 높였다. 그리고 3층에서는 각 공간이 외부 공간과 모두 연결되어 시각적 확장을 고려 하였기에 공간의 다양함이 특징적으로 나타나게 되었다.

Q. 일반적으로 협소주택은 주요 동선이 수직적으로 이루어지는 경우가 많다. 이러한 수직적 생활에서 오는 사용자의 불편함을 극복하기 위한 대안은 무엇인가?
A. 협소주택이 수직적인 구성이 되야 하는 조건이지만 3층과 다락의 수직적 연결 공간을 최대한 계획하였다. 그 연결 공간으로 중정, 내부 계단, 외부 테라스인데 각 부분을 연결함으로써 층으로 인해 단절된 공간의 한계를 공간적으로 풀게 되었고 계획한 수직 공간들로 인해 채광과 환기가 되어 건강한 삶의 공간으로 완성되었다.

Seongsan-dong mix-use housing

Q. What do you think "small houses" are?
A. Small houses are highly likely to be narrow physically, but I'd like to define them based on a simple life. I think small houses should be places where one lives a simple life, organizing complicated and uncoordinated life, rather than one's life should be simple in a house because the house is small.

Q. What were your client's requirements and what was your architectural approach for them?
A. Although it was a small house on a small plot of land, the client asked to include a store that would make some profit. The store was planned for the first story, but if the story height was higher, the area of the highest floor of the house could be reduced a lot due to the setback line limit for daylight. For this reason, the ground height of the first floor was lowered from the road by 1 m, and thus high volume could be obtained even though the area was small, and the area of the second and third floors was planned to be a little broader.

Q. What were your ideas for the creation of maximum space within a small site?
A. First, a big wooden louver door was created for the store on the first floor on the aspect which contacted the road to secure a view from the road and the store. In the dwelling space, the living room, kitchen and dining room were planned in one space on the second floor, but making ground levels different considering the height of furniture increased the use of space. And the third floor is characterized by a diversity of space because visual expansion was designed by connecting each space to outdoor space.

Q. Generally, the major circulation of small houses is made vertically. What is an alternative to overcome the inconvenience to users caused by this?
A. The small house was requested to be composed vertically, but I planned a vertical connecting place between the third floor and the attic as best as I could. The connecting space is composed of a courtyard, internal staircase, and outdoor terrace. By connecting them all, the limitation of space severed by stories was also solved spatially. It could also get ventilation and lighting due to the planned vertical spaces and thus was completed as a living space for health.

35m²

Triangular Balcony Renovation

Location
Melbourne, Australia
Use
House
Built Area
9m²(B1), 35m²(GF), 45m²(1F), 45m²(2F)
Total Floor Area
134m²

Floor
B1 - 3F
Exterior Finish
Aluminium standing seam, matt black, paint
Interior Finish
Birch plywood, mirotone clear satin gloss seal

Project Architect
Fooi-Ling Khoo, David Brand
Photographer
Nic Granleese

Acute house

삼각형 모양의 대지를 활용한 어큐트 하우스

OOF! architecture

$25m^2 - 35m^2$

BEFORE

Acute house

> 독특한 대지 조건을 이용하고자,
> 일반적인 가족용 주택의 구조와
> 생활방식에 변화를 주었다.
> 기본적인 공간 마련을 위해
> 다층구조를 도입해 공간을 구성하고,
> 계단실의 수직 공간을 이용해
> 시각적으로 사생활을 보호하는
> 공간분리를 이루었다.

어큐트 하우스(Acute House)는 건축가에게는 악몽과도 같은 난제를 21세기형 소형 가족 주택으로 탈바꿈한 프로젝트이다. 좁다란 예각 삼각형 대지와 보존 지구의 까다로운 규정 등 제약사항을 역으로 최대한 활용해 뾰족한 쐐기 형태의 새 주택으로 설계했다.

비막이 판자로 지었던 기존의 빅토리아 양식 주택은 극도로 낡아 주거가 불가능한 상태였지만 이웃과 새 소유주 등 모두에게 사랑받는 곳이었다. 우리는 기존 판자집의 특성을 보존하기 위해 뒤틀린 비막이 판자와 울타리 말뚝 외에도 문 손잡이, 환기구, 주소판 등 기존 주택을 이루던 자재와 각종 부품을 최대한 재활용했다. 이 자재들은 섬세한 박물관 유물을 나누듯 조심스럽게 떼어내 라벨을 붙여 보관해 두었다가 원래 위치에 다시 설치했다. 이렇게 재설치한 과거의 요소는 신축된 건물과 대조를 이루며 더욱 돋보였고 원래의 집이 새로운 모습으로 그 역사를 계속 이어갈 수 있게 만들었다.

전통 + 환경적 맥락

완공된 쐐기 모양의 주택은 독특하면서도 주변 지역의 특성에 부합하는 설계를 갖추었으며 부지의 조건과 이에 따른 맥락적 환경에 의한 제약과 기회를 모두 잘 활용한 반응형 건축물이다. 해당 지역의 전체적인 맥락과 '운율'을 맞추면서도 세부적으로는 극도로 현대적인 디자인을 통해 구제할 수 없는 상태였던 이 지역의 주요 랜드마크를 보존해냈다.

가상 정원

부지의 환경적 제약과 형태로 인해 용적률 100%의 설계를 할 수 있도록 지역 의회로부터 허가를 받았지만 그 반대급부로 옥외 공간이 0%가 되었다. 그 결과 가족 생활에 필요한 실내 공간을 꾸려내면서 야외 공간을 즐길 방안도 마련해야 했다. 정원의 부재는 잔디와 같은 초록빛 카펫을 깐 계단실, 공중 화분, 수초와 물고기가 있는 중앙 수족관 등 인공적인 실내 조경 및 채광 및 전망이 좋은 설계를 통해 상쇄했다. 천장 높이의 미닫이 문과 차양막을 열면 거실이 베란다처럼 트인 공간으로 변모하고 발코니도 뾰족한 형태지만 예상외로 넓어 마치 길 위에 놓인 요트 같은 분위기를 더한다.

**In order to take advantage of unique site conditions, we have changed the structure and lifestyle of typical family homes.
In order to provide a basic space, we constructed a space by introducing a multi-layered structure, and achieved a spatial separation that protects privacy visually by using the vertical space of the staircase.**

Acute House is the transformation into a compact 21st century family home. The severe limitations of a tiny, very triangular site and the demanding heritage context have resulted in a pointy new wedge of house that is designed to exploit its problems.
The original, and extremely decrepit, Victorian weatherboard cottage had become impossible to inhabit but was well loved by the neighbourhood as well as its new owners.
We tried to retain its weathered character by re-using as much original fabric as possible from warped weatherboards and fence palings to random accumulations such as door knobs, vents and street numbers. Like fragile museum artefacts, these were carefully removed, labelled, stored and re-installed in their original location on a new mount that not only highlights their charms by contrast but allows the house to live again in a new way.

Heritage-ous-ness + Context
The resulting new wedge of house is designed as an unusual but highly responsive approach to the character of the surrounding neighbourhood, and to the challenges and opportunities for responsive architecture presented by the site and its immediate context. It takes on the challenge of preserving an important but almost unsalvageable local landmark by working within the general typology of the surrounding neighbourhood, "rhyming" with its housing stock while remaining resolutely contemporary in its expression and articulation.

Virtual Gardening
While site area limitations and geometry allowed the council to permit building over 100% of the site this advantage came with the counterbalancing disadvantage of 0% outdoor space. As a result, the house interiors had to accommodate the needs of a family as well as providing them with the enjoyments of the great outdoors. This total lack of garden is offset by the artificial internal landscape of the stairwell with lawn green carpets, hanging plants, a central aquarium of aquatic plants and fish and a sunny outlook to every room. Full height sliding doors and screens open up the main living level as a virtual verandah and the pointy, but surprisingly generous, balcony provides the ambience of a yacht in the street.

25m² - 35m²

SECTION

1 BALCONY
2 ENTRY LOBBY
3 HOME OFFICE/ GUEST ROOM
4 KITCHEN
5 DINING
6 STAIRWELL
7 POWDER ROOM
8 BALCONY
9 WINDOW SEAT
10 MASTER BEDROOM
11 ENSUITE

25m² - 35m²

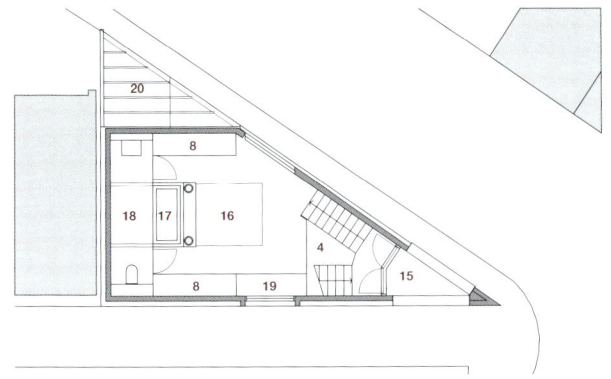

2ND FLOOR PLAN

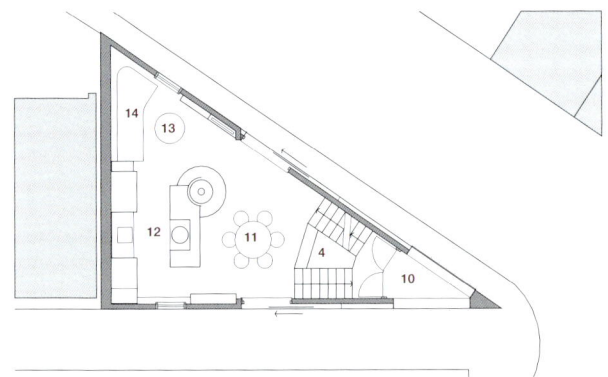

1ST FLOOR PLAN

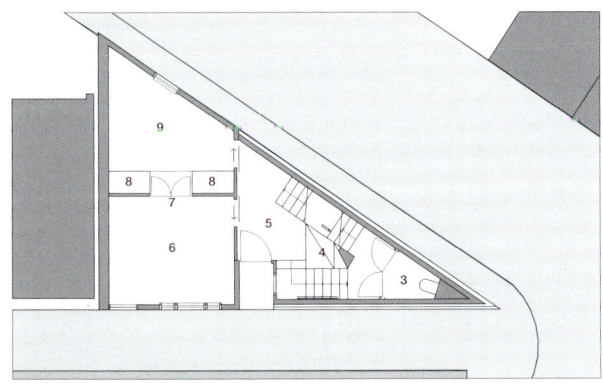

GROUND FLOOR PLAN

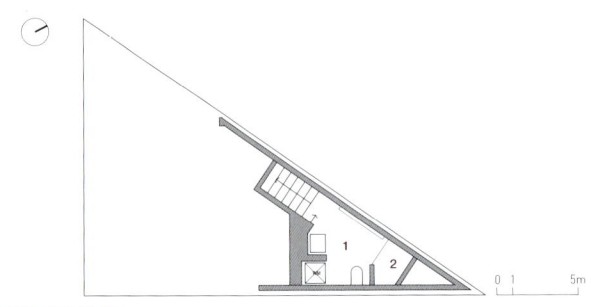

B1 FLOOR PLAN

Acute house

1. BATHROOM+LAUNDRY
2. SHOWER
3. POWDER ROOM
4. STAIRWELL
5. ENTRY LOBBY
6. HOME OFFICE/GUEST ROOM
7. CONNECTING DOOR
8. ROBES
9. KID'S BEDROOM
10. BALCONY
11. DINING
12. KITCHEN
13. LIVING
14. BUILT-IN SOFA
15. BALCONY BELOW
16. MASTER BEDROOM
17. BATH
18. OPEN SHOWER
19. WINDOW SEAT
20. SERVICES

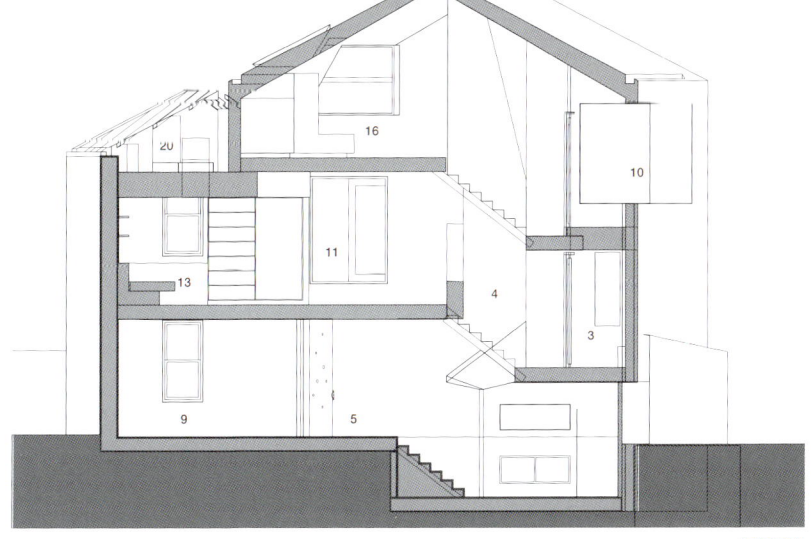

SECTION

$25m^2 - 35m^2$

OOF! architecture

interview

'작을수록 좋다'는 원칙 설계 적용.

공간 마련을 위해 다층 구조를 도입하고, 계단실의 수직 공간을 이용해 시각적으로 사생활을 보호하며 공간을 분리하였다.

Q. 협소하고 제한적인 대지환경에서 최대한의 공간을 만들어 내는 아이디어는 무엇인가?
A. 이렇게 독특한 대지 조건을 기회로 삼으려면 일반적인 가족용 주택의 구조와 생활방식에 변화를 주어야 했다. 가족 생활에 필요한 기본적인 공간을 마련하기 위해 다층 구조를 도입했으며 이를 부지의 형태에 적합하게 배치했다. 공간을 여러 층으로 구성한 뒤 계단실의 수직 공간을 이용해 시각적으로 사생활을 보호하고 공간을 분리했다. 덕분에 소중한 공간을 내벽, 복도 또는 문 등에 낭비하지 않을 수 있었다. 각 층은 막다른 길 없이 동선이 순환하도록 연결해 시각적으로나 물리적으로 답답하지 않고 군더더기 없이 넓어 보이게 했다.

Q. 건물에 지속가능성을 부여하기 위해 전체적 또는 세부적으로 어떤 요소를 사용했는가?
A. 기본적으로 '작을수록 좋다'는 원칙을 기준으로 설계했다. 이 부지에는 구체적인 요건 사항이 있었지만 우리는 언제나처럼 주어진 공간을 다용도로 사용할 수 있도록 명확하고 군더더기 없는 설계를 하고자 노력했다.
도시적 차원이라는 거시적인 관점에서 봤을 때, 전반적으로 조밀해지는 도시의 내부 구조상 이곳처럼 애매하게 남는 대지를 최대한 활용해야 한다. 이런 특이한 대지는 물론 그만큼의 제약을 동반하지만 일반적인 부지에서는 기대할 수 없는 저만의 개성을 지닌 틈새 주택을 세울 수 있어 매력적이다.

비교적 좁은 실내 면적과 개구부의 배치 및 개방형 계단실 덕분에 사계절 내내 아침부터 저녁까지 자연 채광이 이루어지고 한겨울에도 빛이 충분히 들어온다. 건물 북측 개구부에는 그물식 차양막과 블라인드를 시공해 필요 시 여닫을 수 있도록 만들었다. 마찬가지로 각층에 좁은 바닥판을 시공해 층간 통풍이 자연스럽게 이루어지게 했다. 발코니 문과 최상층의 천창은 더운 공기가 개방형 계단실을 타고 반지하층에서부터 꼭대기층까지 순환할 수 있도록 돕는다.
주어진 부지의 규모나 보존 구역상의 제약 때문에 물탱크나 태양열 패널을 설치할 수 없었지만 가스진공관 태양열온수기나 단열판을 적용한 온수바닥난방 등 이용 가능한 곳에 고효율 설비를 시공했다. 모든 벽과 지붕은 단열재로 꼼꼼히 마감했고 모든 창은 목재 창틀에 2중 유리로 밀폐처리했다. 또한 주택 전체에 고효율 하수 및 수도시설을 설치했다.
실내 마감은 구하기 쉬운 친환경 제품을 꼭 필요한 만큼만 소량 사용했으며 조림지에서 생산된 호주산 견목, 합판, 재활용 목제 마감재 등에 페인트칠을 최소한으로 하는 식이었다. 외장재는 관리를 최소한으로 해도 되는 분말 코팅된 알루미늄이나 그냥 두면 자연스럽게 낡아가는 재활용 목제 마감재를 이용했다.

Acute house

Q. What are your ideas to utilize the space on small and unusual site?

A. To take advantage of the opportunities of such an unusual site, the geometry required an adjustment to the layout and lifestyle expectations of a conventional family house. Multiple levels were required to accommodate the basic space needs of a family home and these were accommodated where site geometry best suited them. These spaces are distributed over split levels with the vertical space of the stairwell providing visual privacy and a sense of definition without wasting precious space on internal walls, corridors or doors. Continuous circulation is provided through each floor with no dead-ends, allowing spaces to be kept lean yet feeling spacious and un clogged – visually or physically.

Q. What are the elements of the project contributing towards sustainability (in all its forms)?

A. As a general principle, smaller is better. While there were specific demands on this site, we always endeavour to achieve clear, lean plans that offer multiple overlapping uses of available space.

On a bigger picture, city wide level, the sympathetic densification of our inner urban fabric demands the use of as many of these weird "leftover" sites as possible. While their peculiarities will, of course, present special challenges, they are much more than merely serviceable as they offer unique opportunities for infill housing with enjoyments unachievable on conventional sites.

The comparatively narrow floor plate, arrangement of openings and open stairwell of the house allows natural daylighting throughout the day and evening in all seasons with pleasant sun penetration even in mid-winter. Sliding mesh shading screens and blinds are provided to the north façade openings for the times in the year when this needs to be controlled. Likewise, the narrow floor plate allows natural cross ventilation within each floor. The balcony doors and openable skylights at the top of the roof vent warm air from half basement to top floor via the open stairwell.

While it was not possible to incorporate water tanks or photovoltaic panels due to site and heritage constraints, high efficiency appliances such as a gas boosted evacuated tube solar hot water system and insulated slab hydronic floor heating are used where possible. All walls and roofs are highly insulated and all windows are timber framed double glazed and draft sealed. High water efficiency sanitary and tapware is used throughout.

The internal finishes palette is minimal and modest using commonly available sustainable materials such as plantation grown Australian hardwoods, plywood and recycled timber cladding with minimal painting. External finishes require either minimal maintenance – powdercoated aluminium – or will be left to naturally weather and deteriorate – recycled timber cladding.

28m²

Rooftop Terrace Sliding door

Location
Mexico City, Mexico
Use
House
Site Area
60m²
Built Area
28m²
Total Floor Area
60m²

Exterior Finish
Cement stucco, white paint, exposed brick
Interior Finish
White paint

Construction
PALMA
Photographer
Luis Young

Narvarte Terrace

옥상공간을 활용한 소형주거
나바르테 테라스

PALMA

AFTER

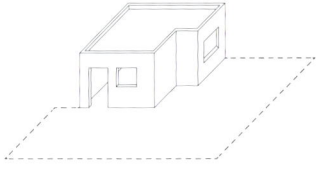

작은 방을 확장해 개별 출입문이 있는 독립된 원룸 형태의 공간으로 활용하고자 했다. 벽 하나를 부분적으로 철거해 추가된 공간에 작은 주방, 수납장, 소파침대를 설치했다. 두 공간은 금속제 지붕으로 연결하고 이 지붕은 처마처럼 바깥쪽까지 뻗어 그 아래로 지붕이 있는 테라스 공간을 형성한다.

멕시코시티의 주거 밀도가 높아지고 시내 일부 지역에서 주택 수요가 증가함에 따라 주민들은 옥상을 최대한 활용할 수 있는 방법을 찾기 시작했다. 이번 프로젝트에서 클라이언트는 작은 방을 확장해 개별 출입문이 있는 독립된 원룸 형태의 공간으로 활용하고자 했다. 벽 하나를 부분적으로 철거해 추가된 공간에 작은 주방, 수납장, 소파침대를 설치했다. 두 공간은 금속제 지붕으로 연결하고 이 지붕은 처마처럼 바깥쪽까지 뻗어 그 아래로 시붕이 있는 테라스 공간을 형성한다. 이중벽과 포켓도어 구조의 미닫이문으로 실내 공간이 테라스를 향해 완전히 열릴 수 있다. 기존 구조에서의 절단면 부분은 시멘트 치장벽토와 백색 페인트로 마감해 강조했고 원래 있던 벽돌 부분은 그대로 노출시켰다.

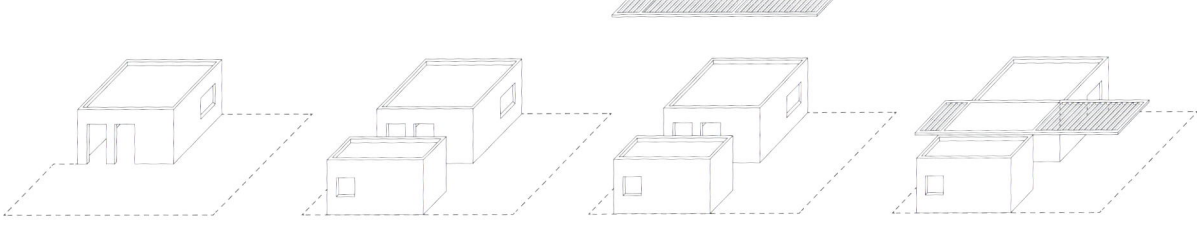

AXONOMETRIC

$25m^2 - 35m^2$

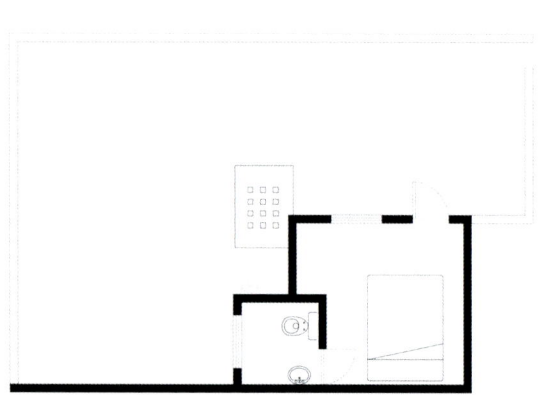

PLAN BEFORE

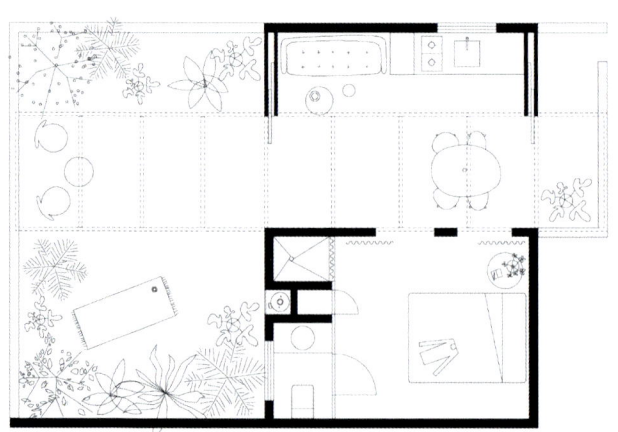

PLAN AFTER

The client was looking to expand a small extra room to be used as a full studio with an independent access. One of the walls was partially demolished in order to add a second volume which would house the kitchenette, storage space, and a sofa-bed. The two volu¬mes are connected by a metal roof which extends out to the edges of the house to create a covered patio.

In Mexico City, the growing demand for housing in certain areas, as well as the increase in density, has motivated the population to find ways of making the most out of their rooftops. In this case, the client was looking to expand a small extra room to be used as a full studio with an independent access. One of the walls was partially demolished in order to add a second volume which would house the kitchenette, storage space, and a sofa-bed. The two volumes are connected by a metal roof which extends out to the edges of the house to create a covered patio. Thanks to a double wall and sliding pocket doors, the studio can be open up completely towards the terrace. The 'incisions' made to the existing structure are highlighted with a cement stucco finish and white paint, leaving the older brick exposed.

25m² - 35m²

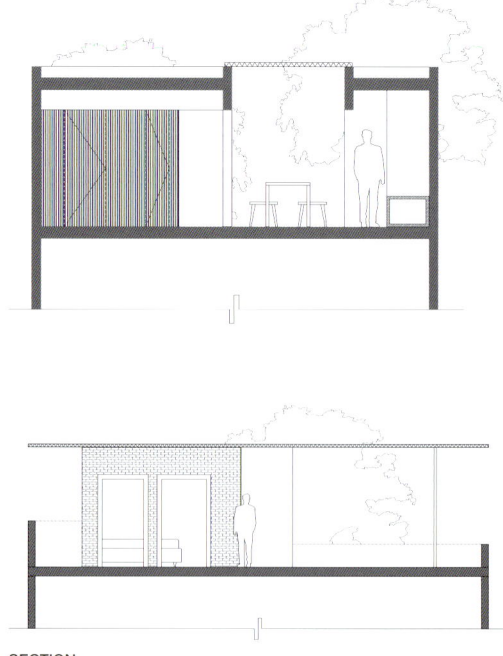

SECTION

PALMA

interview

실내 면적은 28m²에 불과하지만 미닫이문을 열어 옥외 공간과 연결하면 탁 트인 공간감과 함께 거의 두 배에 가까운 공간을 활용할 수 있다.

Q. 클라이언트의 요구사항은 무엇이었으며, 그에 따른 건축적 대응은 어떤 것이었나?
A. 클라이언트는 옥상 공간을 최대한 활용하고 싶어했고, 옥상에는 멕시코시티에서 흔히 볼 수 있는 입주 도우미용 작은 방이 있었다. 우리는 원래 있던 방에 두 가지 요소를 더했다. 첫 번째로 작은 주방과 수납 공간, 소파 베드를 위한 공간을 확장하고, 두 번째로 금속제 지붕을 얹어 기존 공간과 추가된 공간을 잇고 옥외 공간까지 연결해 실내·외의 공간 가치를 더했다.

Q. 협소한 대지 위, 수납공간을 창출하기 위한 아이디어는 어떤 것이었으며, 수직적으로 프로그램 공간구성을 어떻게 하였는가?
A. 우선 휴식을 위해 필요한 공간을 먼저 구획하고 일반 주거 용도를 위한 공간은 그 다음으로 돌렸다. 즉, 기존 공간에 수면, 샤워, 화장실과 같은 개인 주거 공간을 두고, 새로 만든 공간은 주방이나 소파와 같이 보다 개방된 활동을 위한 용도로 만든 후, 금속제 지붕으로 두 공간을 이어 그 사이를 야외로 개방된 식사 공간으로 활용할 수 있도록 했다. 실내 면적은 28m²에 불과하지만 미닫이문을 열어 옥외 공간과 연결하면 탁 트인 공간감과 함께 거의 두 배에 가까운 공간을 활용할 수 있다.

Q. 예산을 줄이거나, 제한된 예산 안에서 최대한의 퀄리티를 만들어내기 위한 방법은 무엇이었나?
A. 지나치게 번거로운 과정 없이도 효과를 극대화할 수 있는 요소들을 찾아 가능한 부분은 연결한 후 이에 집중하는 것이 중요하다고 생각한다. 이번 작업에서 벽체의 페인트를 벗겨 드러난 벽돌과 시멘트 치장벽토로 새롭게 마감한 부분을 대조적으로 조화시킨 것도 그런 예로 볼 수 있겠다. 시공업체도 석조, 금속, 목공 작업과 같이 꼭 필요한 부분으로만 한정해서 시간과 비용을 최소화해 정해진 예산 내에서 공사를 마칠 수 있도록 했다. 목재는 저렴한 소나무를 이용하되 디테일을 살려 벽돌과 백색 치장벽토 마감에 따스한 분위기를 더할 수 있었다.

Narvarte terrace

Q. What were your client's requirements and your architectural approach for them?
A. The clients wanted to make the most out of their rooftop which had a pre-existing small room for the house employee, this kind of rooms used to be very common in Mexico city. We decided to make two main additions to this pre-existing space. The first is a new volume which houses the kitchenette, storage space and a sofa bed. The second one is a metal roof which connects the pre-existing volume and the new one and also integrates the outside space and lets it in adding value to the unit.

Q. Within a limited construction budget, would you be able to share your tips to make a higher quality project?
A. We think it is important to synthesize and concentrate on a few elements which will bring life to the project without much hassle, for example we decided to give contrast by peeling the paint and showing the brick on the pre-existing volume, and then highlighted the new parts with a cement stucco finish. Also we tried to only use a few contractors (masonry, metal work, carpentry) so it wouldn't take too much time and we were able to keep the budget under control by doing that. For the woodwork we used pine wood which is very cheap to get but tried to give some detail to it, so it could bring warmer touches to the brick and white stucco finishes.

Q. What were your ideas to utilize the space & storage on such a small site and how to zone the program in a small area?
A. A. We listed what were the main things you do while you are on vacation, and decided to build the program around this leaving behind some requirements you normally have for more permanent living spaces. With that on mind the preexisting volume houses the private activities (sleep, shower, bathroom) and the new one the more public ones (kitchen and sofa), the metal roof connects this two and covers a casual dining room that opens to the exterior area. Although the interior area has only 28 m2 you can open the sliding doors and let the outside in which gives you a much more spacious and rich spatiality, almost doubling the area

CIRCULATE HOUSE 258p

S1927 272p

HAT 280p

H4912 288p

SUMMERHOUSE T 298p

Y HOUSE 308p

VICOLO 318p

SUBAKO 332p

35m² - 41m²

41.82m²

Skipfloor Rooftop Terrace

Location
Tokyo, Japan
Use
House
Site Area
70.05m²
Built Area
41.82m²
Total Floor Area
95.65m²

Floor
B1 - 2F
Exterior Finish
Lysin spraying
Interior Finish
Emulsion paint finish

Project Architect
Tsuyoshi Kobayashi
Construction
Show-Yo
Photographer
Koichi Torimura

Circulate House

공간의 연속과 분할이 조화를 이루는 순환가옥

another APARTMENT

$35m^2 - 41m^2$

Circulate house

실내를 스킵플로어 구조로 구성하고 각 층을 마치 한 획의 붓질로 그은 듯 공간을 가로지르는 계단실로 연결했다. 공간 구조는 단순하지만 지하의 작업실부터 옥상 테라스와 주변 환경까지 단계적으로 연결한 설계를 통해 기분 좋은 연속성과 공간 분할을 동시에 느낄 수 있다.

이 주택은 도쿄 도심 내 면적이 70m^2 정도되는 대지에 자리하고 있다. 대지는 넓지 않지만 주변 주택가에는 근처에 벚꽃나무가 줄지어 서있는 등 나름의 매력이 있다.

클라이언트는 가족과 함께 시간을 보내고 편히 쉴 수 있는 공간, 영상 편집자인 본인을 위한 작업실, 조용하고 차분한 침실, 벚꽃을 감상할 수 있는 테라스, 그리고 무엇보다도 친구들을 여럿 초대할 수 있도록 다양한 용도를 지닌 상자 형태의 단순한 공간을 요청했다.

그의 요청사항을 바탕으로 공간을 융통성 있게 활용하기 위해 우리는 다양한 특징과 성격을 지닌 공간을 여러 개 마련해 시간과 상황에 따라 거리감에 변화를 주었다.

이를 위해 실내를 스킵플로어 구조로 구성하고 각 층을 마치 한 획의 붓질로 그은 듯 공간을 가로지르는 계단실로 연결했다. 공간 구조는 단순하지만 지하의 작업실부터 옥상 테라스와 주변 환경까지 단계적으로 연결한 설계를 통해 기분 좋은 연속성과 공간 분할을 동시에 느낄 수 있다. 완공된 건물은 이용자의 위치와 활동 내용에 따라 공간을 다양하게 활용할 수 있는 집이 되었다.

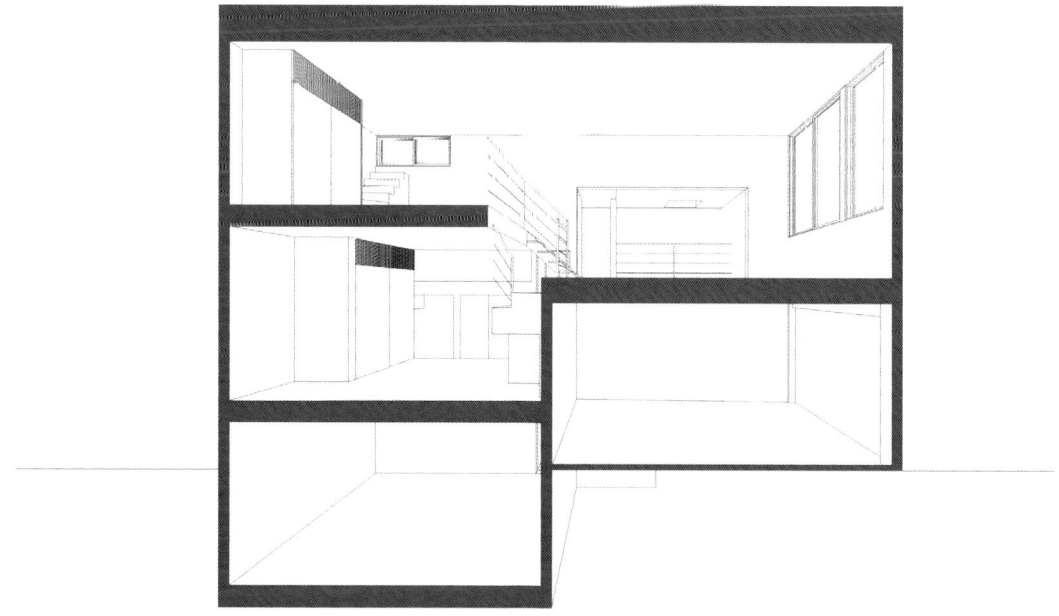

CONCEPT IMAGE

35m² - 41m²

Circulate house

The building is configured with skipped floors, connected by staircases travelling around and crisscrossing as if like a single stroke of a brush. While this configuration has a simple manner, it connects from the atelier space on the basement to the rooftop terrace, and even to the surrounding environment in gradational modes, enabling to feel pleasant continuity and division at the same time.

In order to realize the flexibility that the client requested, we tried to produce a house with multiple spaces in different qualities and natures, altering user's feeling of distance according to the time and conditions.

To do so, the building is configured with skipped floors, connected by staircases travelling around and crisscrossing as if like a single stroke of a brush. While this configuration has a simple manner, it connects from the atelier space on the basement to the rooftop terrace, and even to the surrounding environment in gradational modes, enabling to feel pleasant continuity and division at the same time. The building became a flexible house revealing different aspects, depending on the positions and activities of users.

This residence stands on a site with the area of about 70 sqm in the Tokyo metropolitan area.

While the site is not large, the surrounding residential area has an attractive point such as a row of cherry blossom trees nearby. The requests of the client include a space to spend time and feel at ease with family members, an atelier space for the client as a video editor, a quiet and tranquil bedroom, a terrace to enjoy watching those cherry blossom trees, and above all, a simple, flexible, box-like space allowing to invite many friends to the house.

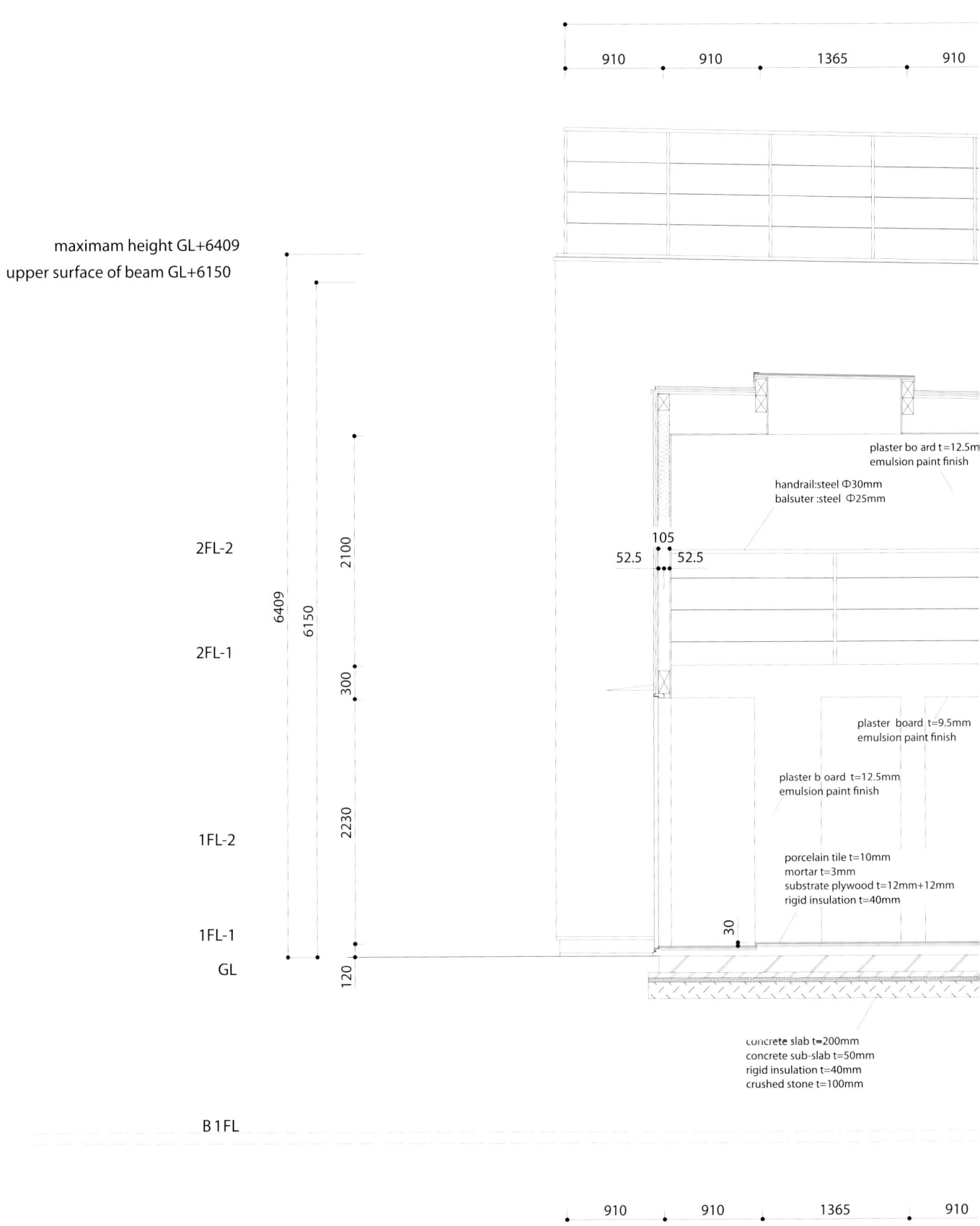

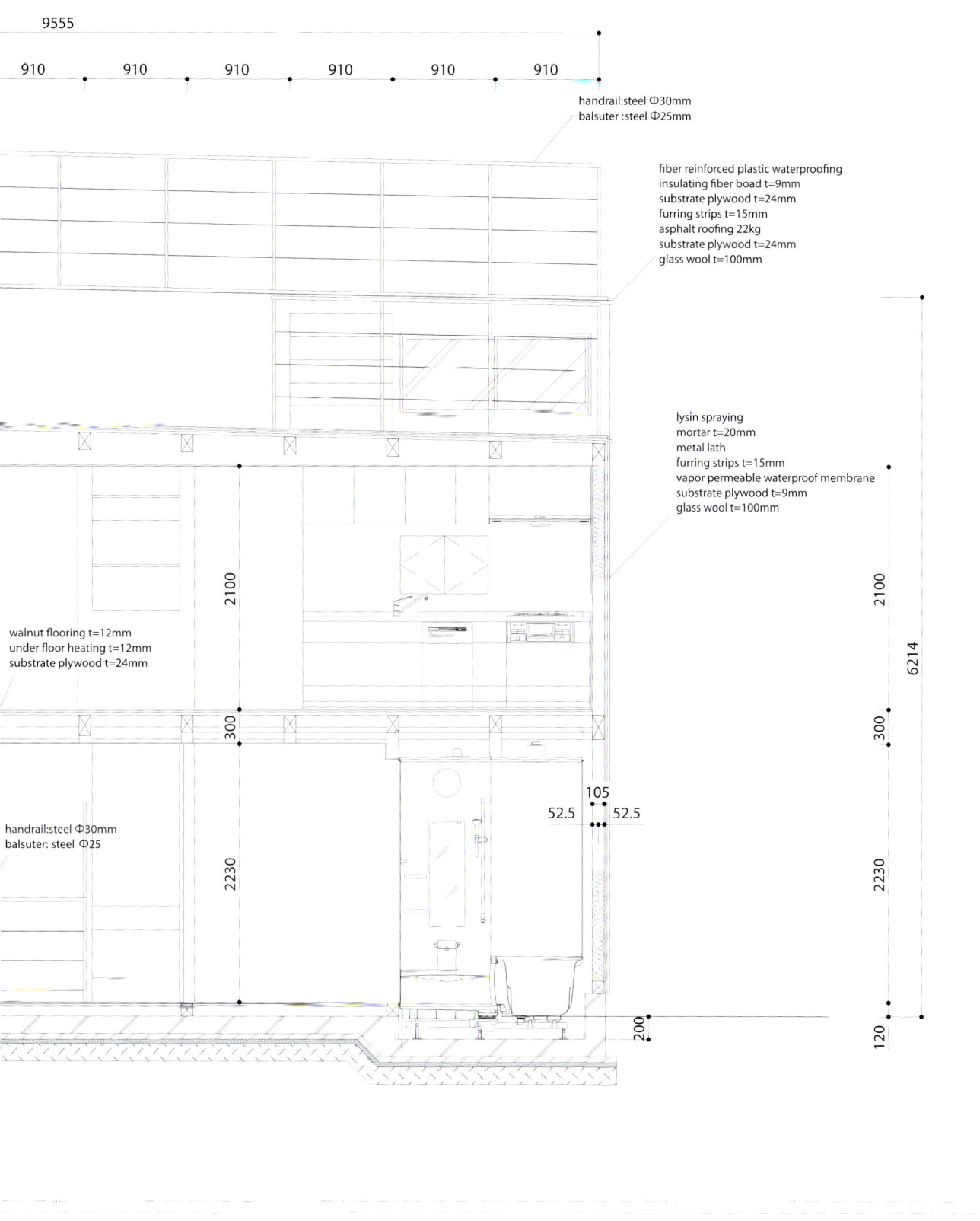

SECTION DETAIL

35m² - 41m²

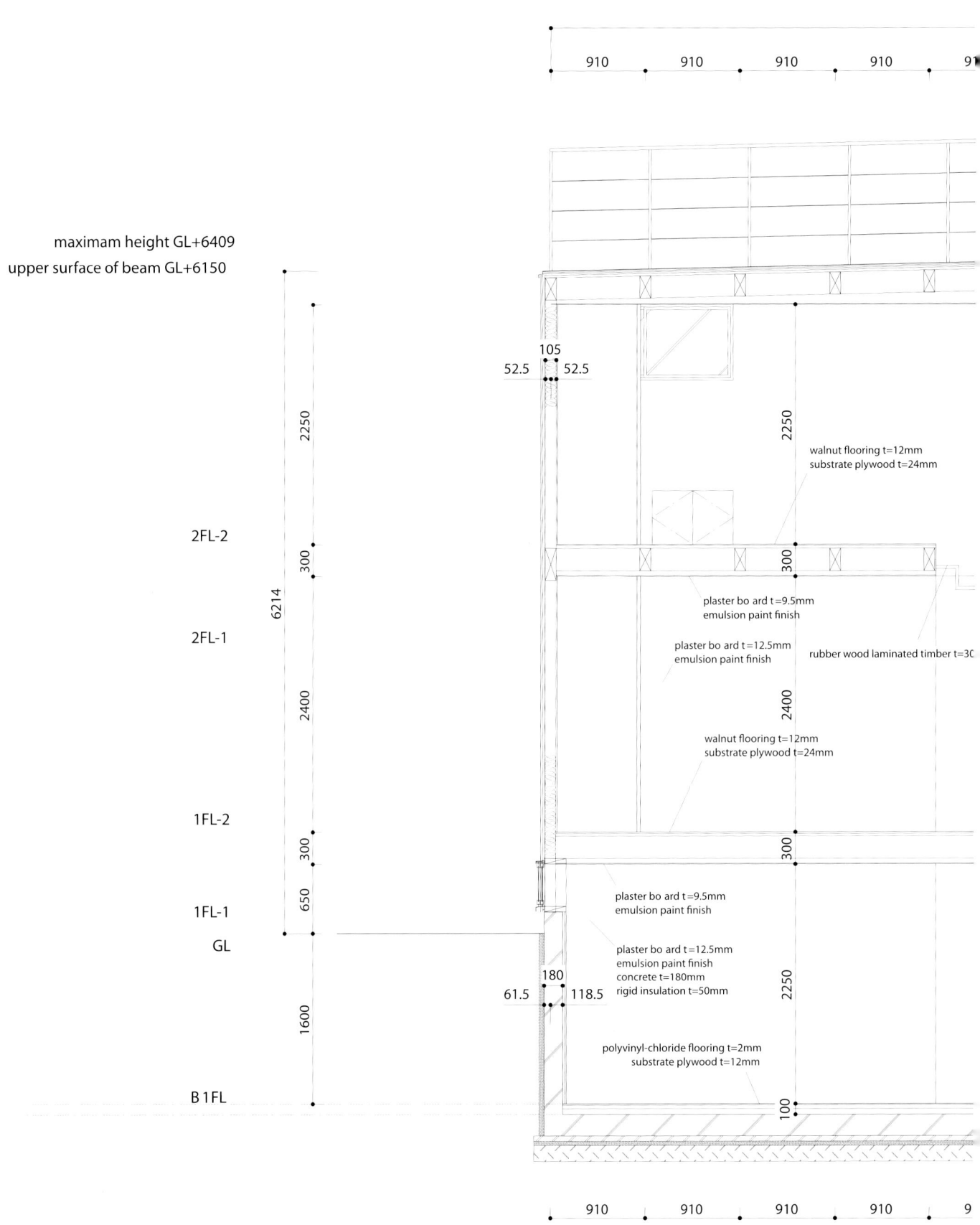

Circulate house

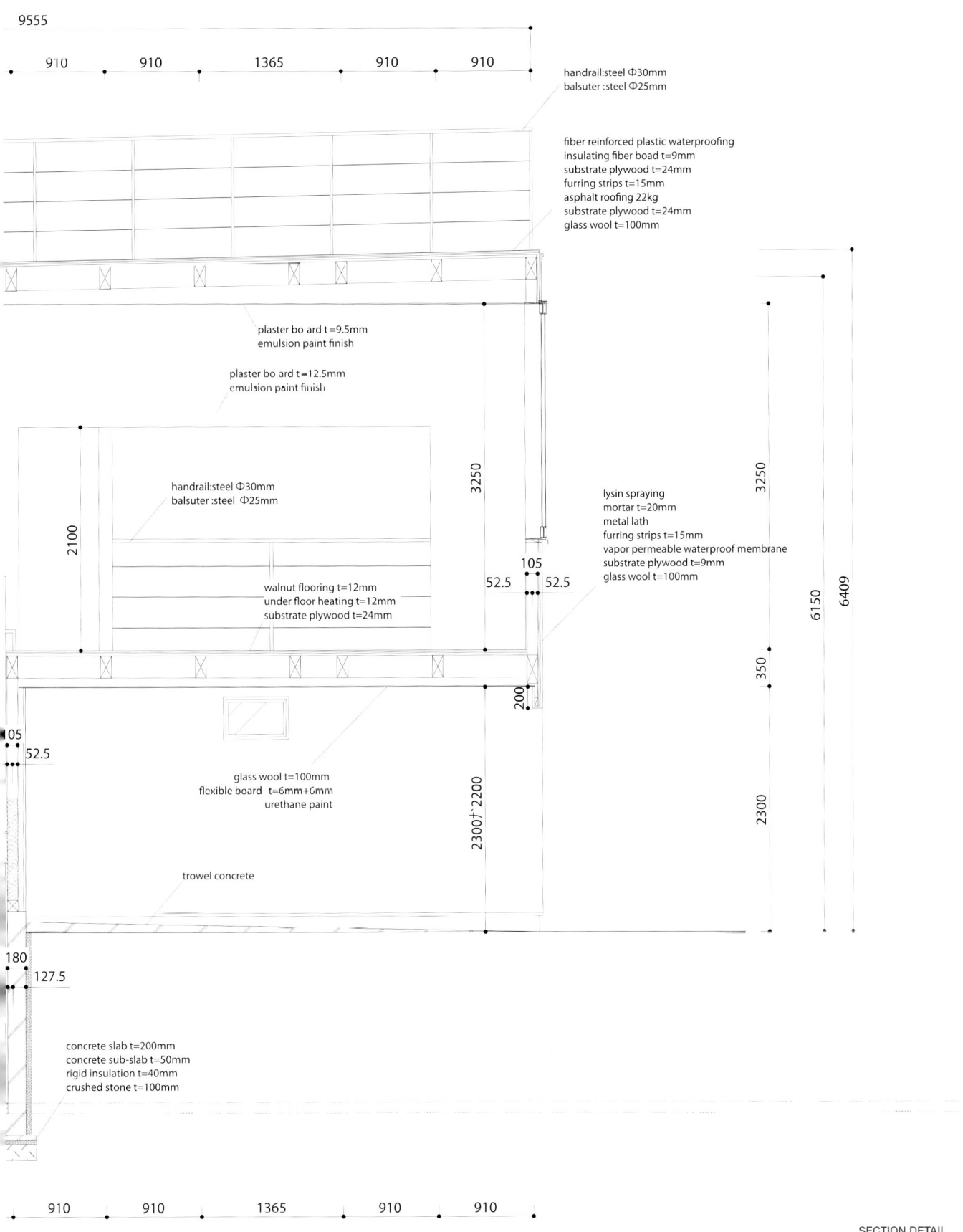

SECTION DETAIL

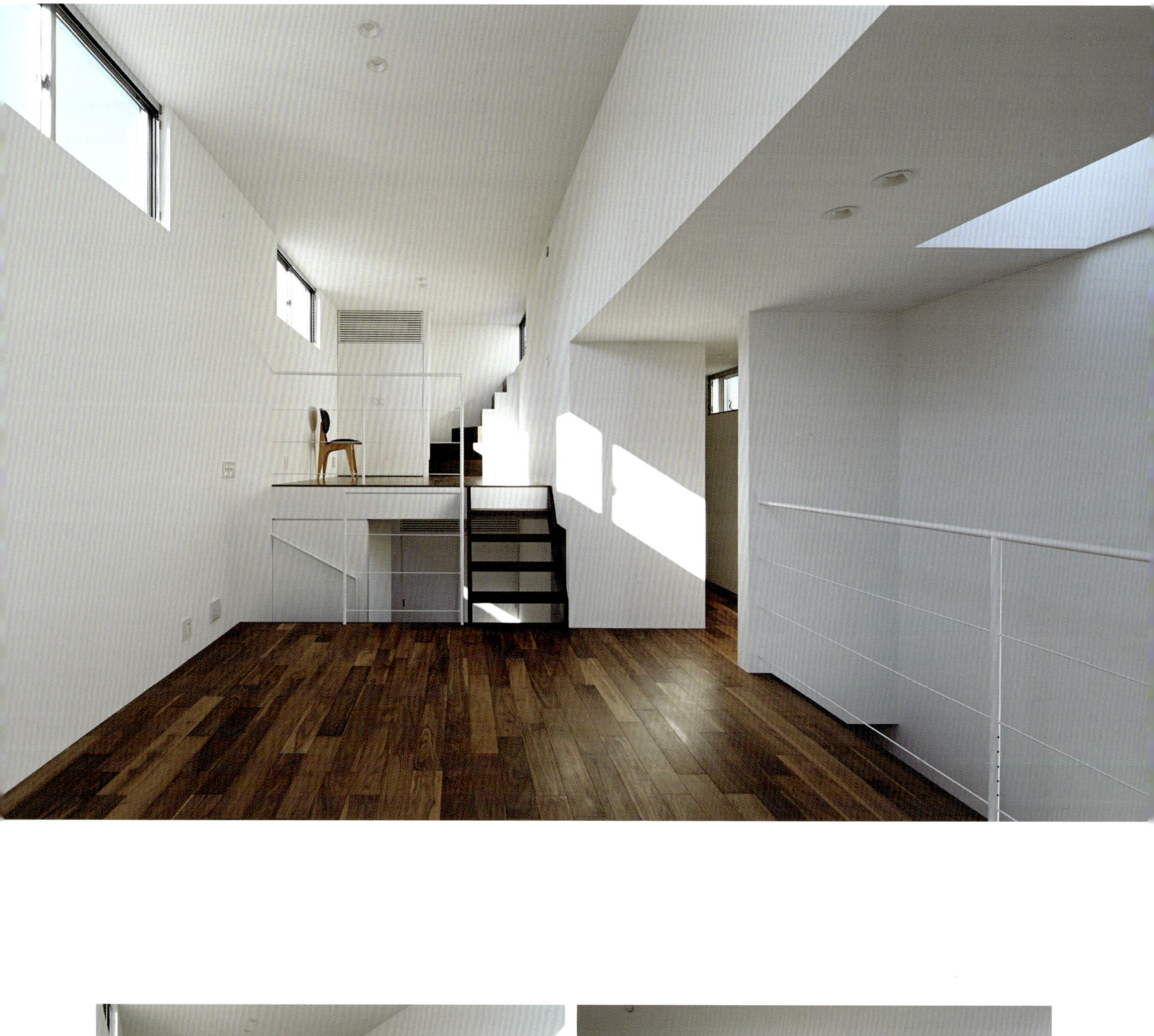

$35m^2 - 41m^2$

Circulate house

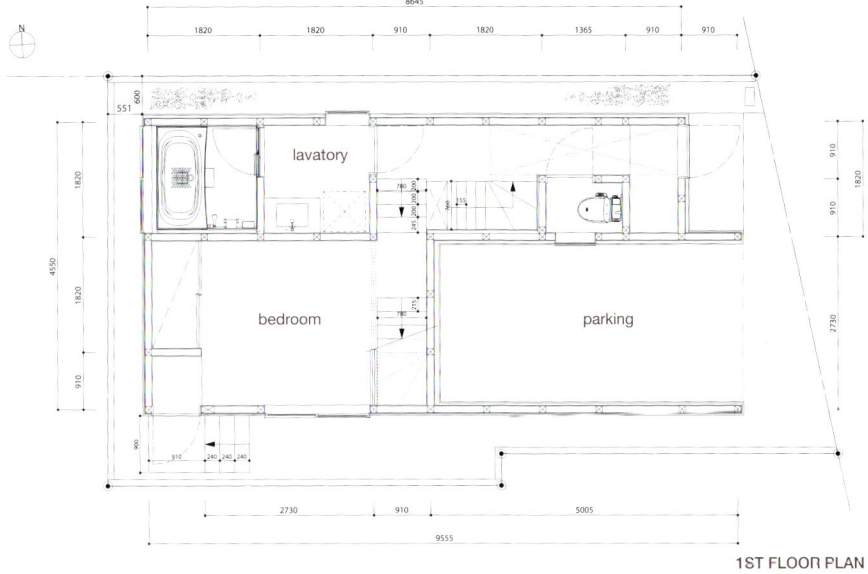

1ST FLOOR PLAN

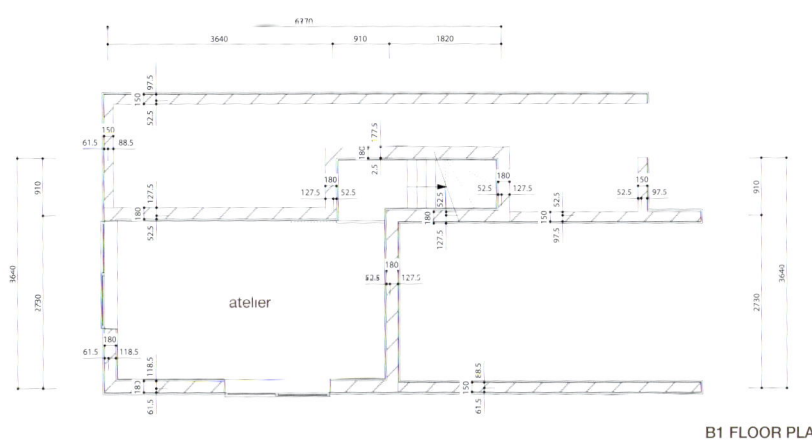

B1 FLOOR PLAN

2ND FLOOR PLAN_2

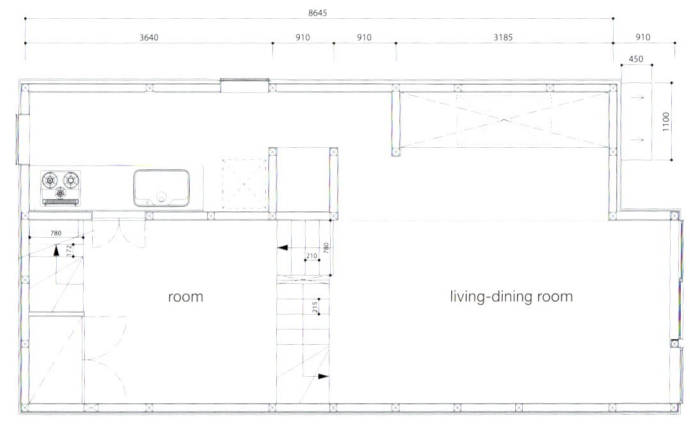

2ND FLOOR PLAN_1

39.60m²

Skipfloor Triangular site

Location
Seoul, South Korea

Use
House & neighbourhood facility

Site Area
71.36m²

Built Area
39.60m²

Total Floor Area
142m²

Floor
5F

Main Structure
Steel reinforced concrete

Exterior Finish
STO(softwood)

Project Architect
Lee Yongeui, Song Kideok

Construction
KINFOLKS STUDIO

Photographer
Byun Jongsuk

S1927

스킵플로어로 공간들을 연결한 적은집 S1927

GONGGAM&KINFOLKS (공감 건축시시무소)

$35m^2 - 41m^2$

건축주의 노후를 위해 1층은
임대료가 나올 수 있는
근린생활시설로 계획하고,
4인 주거에 필요한 각각의 실들을
최소단위로 계획하여 집합시켰다.
중앙 계단을 설치하여 각각의 실들을
연결하는 동선을 만들었으며,
서로 연결되어야 하는 실들을
위해 스킵플로어 형태의 단면을
구성하였다.

석관동

1970년대, 기존의 골목길을 유지한 상태에서 토지구획 정리 사업이 이루어졌다. 이로 인해 석관동 곳곳에는 반듯반듯한 땅의 귀퉁이마다 부정형 필지와 삼각형 필지가 생겨나게 되었다. 이런 삼각형 땅에 지어진 건물들은 집장사들에 의해 무분별하게 지어진 주택들이었고, 특색없는 사선의 골목을 만들어 내고 있다.

석관동 192번지 일원은 석관 제1, 2 주택재개발 지역에 포함되지 않은 곳으로, 자율적인 개발이 가능했다. 이 지역은 경희대학교와 한국예술종합대학교가 인접해 있어 준주거의 수요가 많았기 때문에 많은 저층 단독주택들이 원룸건물이나 빌라로 변하게 되었고, 그 결과 집장사들에 의해 건축되었던 저층 주거지는 4, 5층 빌라와 원룸 건물로 이루어진 무색무취의 빌라촌으로 변해왔다.

삼각형의 필지와 정형화된 공간

개발이 아닌, 시간이 퇴적되어 만들어진 필지는 사각형 모양의 땅이 아니다. 현대에 사는 우리가 영유하는 공간과 오브제는 모두 사각형으로 이루어져 있다. 따라서, 비정형적으로 만들어진 공간은 그 공간을 구성하는 가구의 제작을 포함하여 건축의 모든 공정에서 별도의 주문제작된 것만을 허락하게 되고, 많은 비용을 투자해야 하는 일이 필연적으로 발생하게 된다. 즉, 시간이 만든 자연스러운 땅에서 자신의 집을 지어 소소하게 살고자 하는 사람들에게 땅 모양에 맞춰 지어진 비정형적인 공간은 역설적이게도 어울리지 않게 된다. 積恩集(적은집) S1927에서는 필지와 공간의 거리감을 좁히고 싶었다. 이 거리감을 좁히기 위해서 사각형의 정형화된 공간을 층층이 쌓아 조합하여 긴 삼각형의 필지에 어울리는 평면을 형성하였다.

시스템

부부와 2명의 딸이 함께 거주하는 주택이다. 부부의 노후를 위해 1층은 임대료가 나올 수 있는 근린생활시설로 계획하고, 4인주거에 필요한 각각의 실들을 최소단위로 계획하여 집합시켰다. 중앙 계단을 설치하여 각각의 실들을 연결하는 동선을 만들었으며, 서로 연결되어야하는 실들을 위해 스킵플로어 형태의 단면을 구성하였다.

집합의 건축으로

積恩集 S1927에서는 9개의 개별요소를 집합하여 개별의 건축을 완성하였다. 개별요소는 기능하는 실이고, 개별의 건축은 협소주택이다. 개별의 건축이 모이면, 집합의 건축이 될 것이다. 집합의 건축은 마을이다. 우리가 살아가는 도시의 마을을 꿈꾸는 것은 건축가만의 이상향은 아닐 것이다.

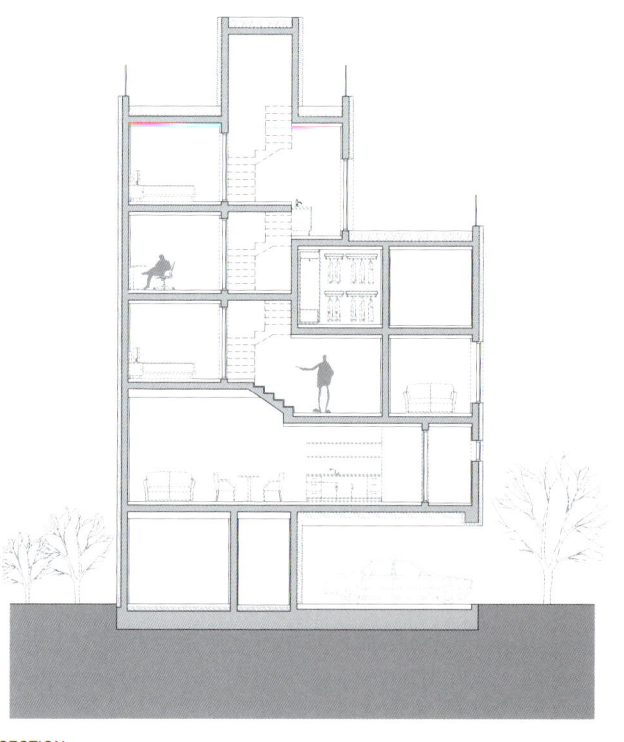

SECTION

This house was designed for a family of four, a father, a mother and their two beautiful daughters. Commercial facilities are located on the ground floor for the couple to earn the rent fee for their after retirement life. All the residential spaces for the family were designed in the smallest units and then assembled. The centralized staircase plays a role as a vernicle and main circulation that connects each units and rooms. Skip-floor type plan is used for the spaces that should be linked to each other.

Sukkwandong

In 1970s, based on the existing streets, there was land rearrangement business led by government. Due to the rearrangement, in SukKwanDong, the land shape of its area became triangular and irregular shape. Most of the buildings built in this period were planned indiscriminately as a profit-oriented development. As a result, diagonal streets and lane ways with no particular characteristics were appeared. SukKwanDong 192 area could be developed without any strict regulation, as it was not within the housing redevelopment zone. Since the area was surrounded by Kyung-Hee University and Korea National University of Arts, semi- residential space was high in demand, encouraging single housing to be turned into studio apartment building to rent the place to student. Eventually, the suburb with weak identity was formed.

Triangular plot and regular shaped space

Plots of land which was shaped as time passes was not formed into a land with a regular geometry. In the contrast, space and objects that we use and see these days are mostly formed in rectangular shapes. Hence, irregular or non-rectangular space are bound to cost more to build as the whole process of construction, including manufacturing furniture, requires special treatments or customized order. That is, ironically, the space shaped irregularly to fit onto the shape of the plot becomes no longer suited for those who want to live a simple life in their own house built on their piece of land naturally shaped as time passed. We took the concern seriously and desired to present a solution. In the project S1927, to minimize the disparity between the land and the floor plan, we have created the space piling up rectangular modules on the triangular site so that the triangular land and the floor plan can be balanced.

System

This house was designed for a family of four, a father, a mother and their two beautiful daughters. Commercial facilities are located on the ground floor for the couple to earn the rent fee for their after retirement life. All the residential spaces for the family were designed in the smallest units and then assembled. The centralized staircase plays a role as a vernicle and main circulation that connects each units and rooms. Skip-floor type plan is used for the spaces that should be linked to each other.

Collective architecture

9 elements were assembled together to form an individual architecture in 'S1927'. Each element is functional rooms and the individual architectural is Micro house. When the individual architecture gathers together, they become collective architecture. Collective architecture means a village. Dreaming of an urban village would not be Utopia only for architect but for everyone.

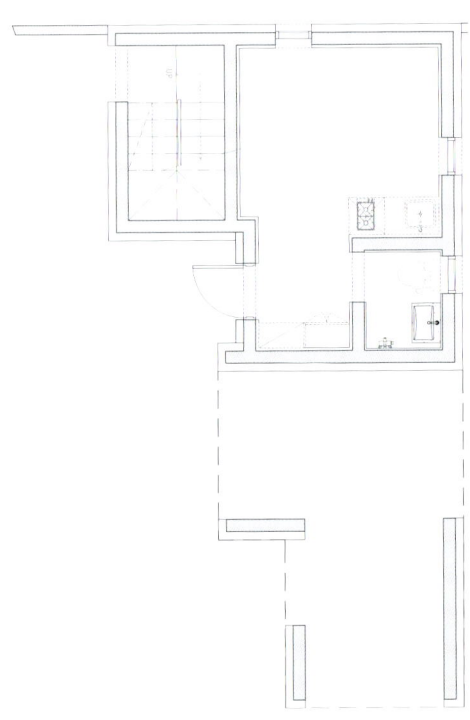
1ST FLOOR PLAN

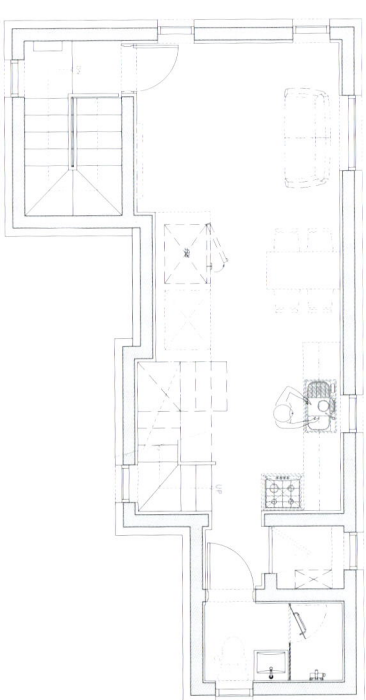
2ND FLOOR PLAN

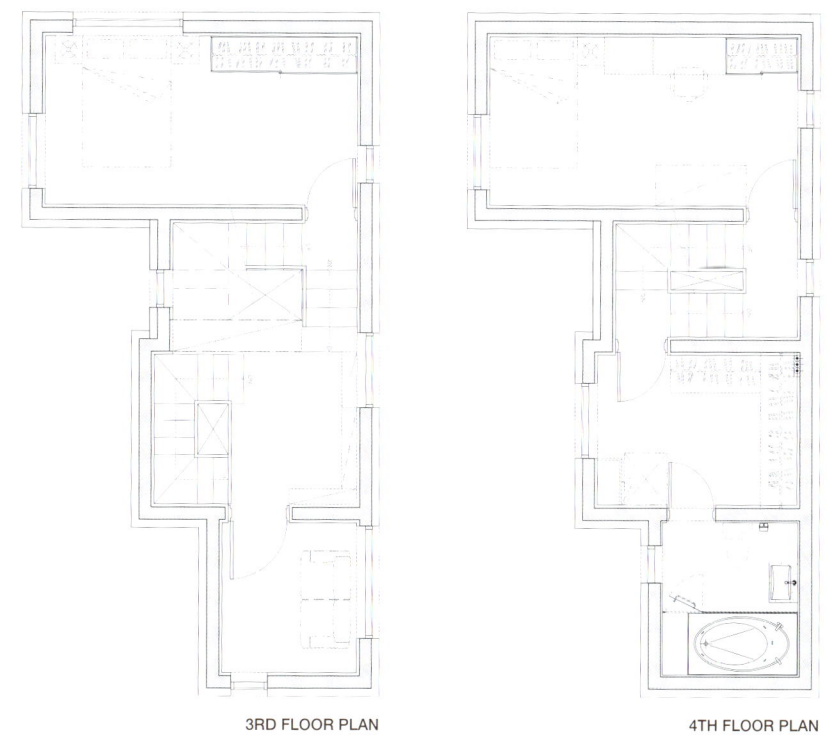

3RD FLOOR PLAN　　　　4TH FLOOR PLAN

39.34m²

Courtyard

Location
Shinjuku, Tokyo, Japan
Use
House
Site Area
66.95m²
Built Area
40.87m²(B1), 39.34m²(1F), 39.34m²(2F)
Total Floor Area
78.68m²

Floor
Basement + 2 floors
Structure
Reinforced concrete, wooden
Exterior Finish
Exposed concrete

Project Architect
Satoshi Kurosaki
Photographer
Masao Nishikawa

HAT

자연광이 풍부하게 들어오는 집 HAT

APOLLO Architects & Associates

$35m^2 - 41m^2$

B1 FLOOR PLAN

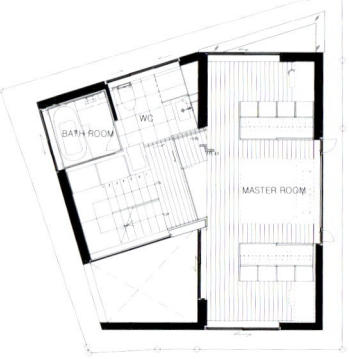

1ST FLOOR PLAN

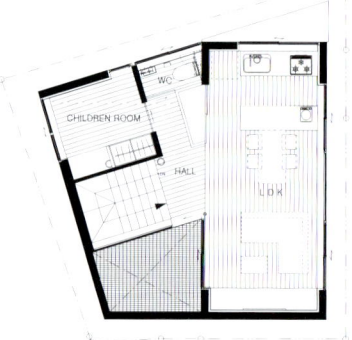

2ND FLOOR PLAN

> 외부에서 보기에는 다소 폐쇄적으로 보이지만, 뜰을 만든 덕분에 내부 공간은 자연광이 풍부하며 개방적인 느낌을 준다.

건축주는 풀타임으로 근무하는 50대 부부이다. 이들의 계획은 어머니와 함께 거주하기 위해 노년 시기에 있는 부모님의 집을 재건축하는 것이었다. 대지는 그리운 옛 도쿄의 흔적이 남아 있으며 인구 밀도가 높은 주거 지역에 위치한다. 이들 부부는 향후 발생 가능한 천재지변에 대처하기 위해 내화성 및 내진성이 높은 강화 콘크리트 내력벽식 구조의 주택 건축을 원했다. 대지 크기가 작은데다가 이형(異形)대지여서 주택 각 면의 후퇴선 경계를 피하면서 최대한의 수용량(收容量)을 확보해야 했다. 주택이 막다른 골목의 맨 끝부분에 지어져서 주택 전경을 볼 수는 없다. 건물은 총 3층 (지상 2층, 지하 1층)으로 구성된다. 지하층의 절반은 지하에 매장되어 있는데, 이는 경계를 완화하여 최대한의 수용량을 얻기 위함이다.

주택의 형태를 잡기 위해 이형대지의 외곽선을 입체적으로 돌출시켰다. 뾰족한 모자를 연상시키는 지붕으로 덮인 콘크리트 박스 구조는 그 매력적인 외관으로 행인들의 눈길을 사로잡는다.
현관의 출입문 옆에 있는 계단으로 내려가면 나오는 반지하층에는 어머니 방이 있다. 전체 평면도는 간결하게 설계됐다. 부부 공간은 1층에 위치하며 남편과 아내를 위한 두 개의 서재와 서재 사이 침실, 욕실 구역이 있다. 외부에서 보기에는 다소 폐쇄적으로 보이지만, 뜰을 만든 덕분에 내부 공간은 자연광이 풍부하며 개방적인 느낌을 준다.
2층에 있는 거실은 후퇴선 경계를 따라 만들어졌다. 지붕틀에 강화 콘크리트 대신 목재 서까래를 사용한 덕분에 복합 구조의 독특한 외관이 탄생했다.
뜰이 있는 이 다락방같은 작은 공간에서는 특별한 구심력을 느낄 수 있는데, 이 공간을 보고 있자면 어쩐지 몽골 유목민들의 주거지인 유르트가 떠오른다. 주변 조명으로 환하게 밝혀진 전체 지붕은 이토록 작은 공간에서 마치 커다란 나무의 보호를 받으며 머무는 것과 같은 안정감을 준다.

$35m^2 - 41m^2$

Although it appears rather closed from the outside, the interior space with a sense of openness with abundant natural light is achieved by providing the courtyard.

The client is a couple in their 50s and both work full-time. Their plan was to rebuild their parents' house in the urban area in order to live with their mother. The site is located in a dense residential area where one can find the remnant of good old days of Tokyo. In order to prepare for possible natural disasters in the future, the couple wished to build a house of reinforced concrete box frame construction with high resistance to fire and earthquakes. Since the site is small with a deformed shape, it was required to achieve the maximum capacity while avoiding setback-line limits on each side of the house.

One cannot have a full view of the house since it is built at the very end of a blind alley. The building consists of three stories, with two floors above ground and a basement floor, and a half of the basement floor is buried underground in order to achieve the maximum capacity by taking advantage of easing of the restrictions.

The outline of the deformed land was extruded three dimensionally to form the house, and the charming appearance of the exposed concrete box topped with a roof resembling a pointed hat catches the eye of passers-by.

Their mother's room is on the semi-basement floor, down the stairs next to the entrance in the entrance court. The entire floor plan is designed compactly: the couple's space is on the first floor with two study rooms, for the husband and wife respectively; a bedroom between the studies; and a wet area. Although it appears rather closed from the outside, the interior space with a sense of openness with abundant natural light is achieved by providing the courtyard.

The family room on the second floor is shaped along the setback-line limits, and the wooden rafters are used for the roof truss instead of reinforced concrete ones, creating a unique appearance of the mixed structure.

One feels a distinct centripetal force in the loft-like small space with a courtyard, which somehow reminding one of a yurt, a dwelling of Mongolian nomads. The entire roof is lit up by the ambient light, creating a sense of security in such a small space, as if staying under the shelter of a big tree.

SECTION

35.10m²

Skipfloor
Roof garden
Various window

Location
Scoul, South Korea
Use
House
Site Area
62.10m²
Built Area
35.10m²
Total Floor Area
110.06m²

Floor
4F
Main Structure
Steel reinforced concrete
Exterior Finish
STO

Project Architect
Lee Yongeui, Choi Yeonjung
Construction
Two&Hand Design
Photographer
Byun JongSuk

H4912

다양한 사이즈의 창들로 풍부한 조망을 제공하는 적은집 H4912

GONGGAM&KINFOLKS (공감 건축사사무소)

35m² - 41m²

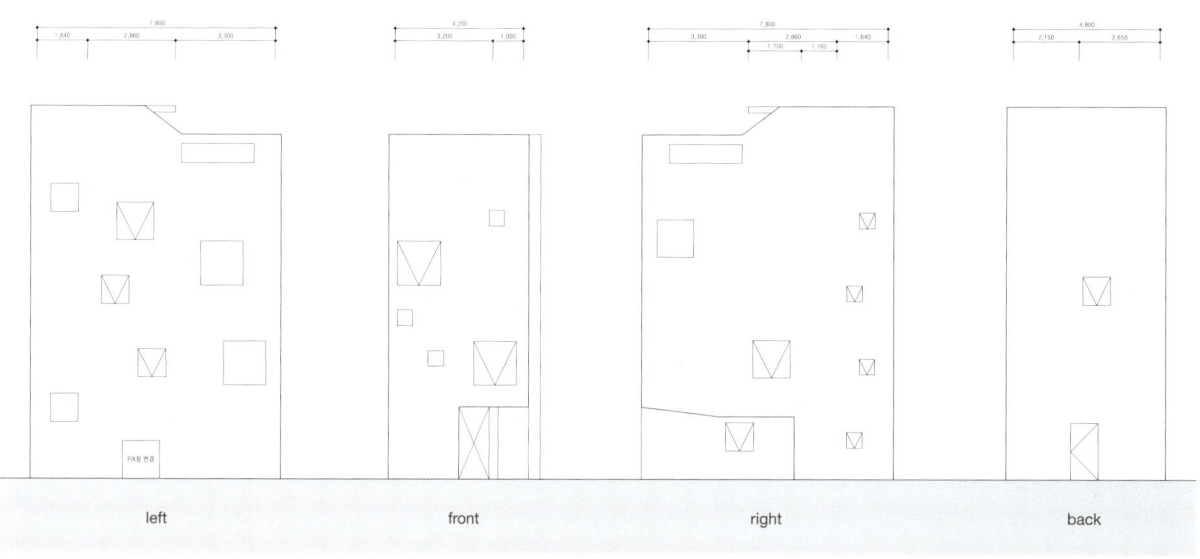

ELEVATION

**공동주택과 같은 발코니와 창은 없지만, 각기 다른 위치에 계획된 다양한 사이즈의 창들은 풍부한 조망을 제공해 줄 것이다.
아파트 단지의 넓은 외부공간은 없지만, 가족만 쓸 수 있는 옥상 정원이 그 공간을 대체해줄 것이다.**

공동주택에 거주한다는 것은

서울에 사는 대다수의 사람들은 공동주택에서 살고 있다. 건축물 용도에 따른 공동주택보다 공동주택이 만들어낸 단위 평면도에 거주한다는 것이 더 맞는 표현일 듯하다. 공동주택이 만들어낸 평면은 단일 모듈 안에 거실과 주방을 기준으로 양 옆의 방을 만드는 형태일 것이다. 다시 방과 방 사이에는 다용도실과 화장실이 만들어져 서로 관계를 맺고 있다.

거실과 주방은 그 집의 중심에 자리하고 있다. 절대적인 힘을 소유한 거실과 주방은 절대 둘 이상 존재할 수 없는 한쌍의 암수와 같은 존재이다. 가족구성원이 생활하는 동선의 중심이며, 감시와 소통을 적절히 조절하는 절대적인 힘을 가지고 있다.

거실과 주방에서 외부 손님을 맞이하게 되면, 가족의 보호는 각 방이 담당하게 된다. 방은 각 개인의 절대적인 쉘터(shelter)로, 그 세대의 모듈로 대변된다.

공동주택은 우리에게 형태의 다양함을 주지는 않지만, 모든 평면에 거실 창이라 불리는 큰 창을 주었다. 그 거실 창은 표준으로 자리 잡았으며, 모든 사람들이 같은 사이즈의 프레임 속 세상을 바라보게 만들었다. 모듈화된 거푸집은 각 층에 다양한 삶의 형태를 담아두기 싫어하다, 각 층은 같은 세대를 수평으로 나열하고, 그것을 수직으로 쌓아 같은 삶 다른 높이를 만들어 주었다.

積恩集(적은집)에 거주한다는 것은

서울이라는 도시에서 단독주택에 거주한다는 것은 쉬운 일이 아니다. 얼마 남지 않은 단독주택지는 고급주거가 아니면, 도시형생활주택과 아파트 재개발로 사라져가고 있다.

18평의 작은 땅에 거주한다는 건 쉬운 일이 아니다. 대지가 일조권과 도로사선의 제한을 이길 힘이 있어야 하며, 상상력이 뛰어난 건축주와 건축가가 필요하고, 공동주택이 만들어낸 시공 기술을 이겨낼 장인도 필요하다.

공동주택과 같은 큰 거실도 없고, 이동하기 불편한 계단과 협소한 화장실도 이용해야 한다. 하지만, 신아가는 농선본 좀 더 다양해질 수 있을 것이다. 1층은 남편의 거실, 2층은 아내의 거실, 3층은 아들의 방, 4층은 부부침실과 다실로 이루어져 적절한 관계를 유지한다. 손님이 방문할지라도 각자의 공간은 자유롭게 생활 할 수 있을 것이다.

공동주택과 같은 발코니와 창은 없지만, 각기 다른 위치에 계획된 다양한 사이즈의 창들은 풍부한 조망을 제공해 줄 것이다. 아파트 단지의 넓은 외부공간은 없지만, 우리가족만 쓸 수 있는 옥상 정원이 그 공간을 대체해줄 것이다.

가족과 삶을 설계하는 것은

공동주택과 단독주택은 삶의 방식이 전혀 다르다. 공동주택은 주어진 환경에 맞춰서 사는 거라면, 단독주택은 가족 구성원이 살고자 하는 환경을 만들어 사는 것이다.

공동주택에서는 살고자 하는 단지에서 원하는 평형대의 동과 호수만 정하면 되지만, 단독주택에서는 동네와 이웃과 문화를 함께 고려하며 결정해야 한다. 또한 가족과 어떻게 살고 싶은지에 대한 상의가 필요하며, 이러한 각각의 공간을 집합시켜주는 설계가 이루어져야 한다.

설계되어진 삶을 선택하는 것과 가족과 함께 설계한 삶을 선택하는 것은 모두가 고민해야 할 숙제일 것이다.

$35m^2 - 41m^2$

> Although there is no balcony or standard size of windows like other typical apartment houses, various size and different location of windows can provide widely opened view and enough natural lights. Instead of having public sharing space like other apartments, private rooftop garden can substitute the need of green space.

Living in apartment housing means

Most of Seoul citizens reside in apartment type housing. Precisely, it is more right to say they live in buildings that are not built to be multi- complex housing at the beginning, yet they are built based on similar unit plans so that more households can fit in in limited size of land. Most of the Plans are modularized having living space and kitchen in center with private rooms on the side like other typical Korean apartment plans. Between the rooms, there are multi-purpose space and washrooms connecting all spaces to be more functional. Livingroom and Kitchen are placed in a center of plan. Strong connection between living space and kitchen are absolute, therefore both of the space are mandatary to be placed near. They positioned in the center of main circulation and it has a power to control natural surveillance and communication among family members. In addition, both of the places are tend to accommodate visitors and friends occasionally and private rooms along with the space keep your room private. Each private room roles as an absolute shelter and each presents a single module representing units. Multi-family housing does not provide diversity to users, but all plans have a large size window. The large window now became standardized and people view outsides in a same window frame no matter where you live. Modularized size of construction forms does not tempt to take various life styles of individuals from all floors.

Each flats rearranged all typical family unit floor plans horizontally and the rearranged floor plans are simply piled up making users having similar attitudes towards life. Only distinctive perspective they could have is that they live in a different floors.

Living in 積恩集 means

Environmentally, building a single house in Seoul is not usual. Nowadays some houses that are not apartments are rebuilt to urban style compact complex buildings or higher apartment buildings. Building a house in an 18 square meter land is quite complicated. Constructing a compact house in a tiny land lot requires strict building standard law such as right of light and setback regulation from road width. Also, outstanding imagination of both architects and land owners and excellent construction skill are necessary. Like other apartment houses, there is not enough space for large living room and in order to have private rooms and washroom putting stairs are mandatory. On the other side, circulation can be more dynamic compare to houses that are planned in a flatland. First floor covers husband's living space, second floor is suitable for wife's living space, third floor is planned for son's private room and lastly fourth floor has a private room for husband and wife and multi-purpose space for every family member. Even though visitors come over, each private areas still remain private. Although there is no balcony or standard size of windows like other typical apartment houses, various size and different location of windows can provide widely opened view and enough natural lights. Instead of having public sharing space like other apartments, private rooftop garden can substitute the need of green space.

Designing plan for your family and life

Living in an apartment house and a single house bring totally different experience to resident. Apartment house is usually planned or built base on certain conditions or in a given space, but single house sets desired environment for particular family. Choosing to reside in an apartment house requires to consider size, location of building and floor level, but living in a single house also necessary to take care for living with neighbors. Also, each family member is necessary to discuss about how their house should be built and how their required spaces are combined and reflected to plans. It is extremely a tough decision to take either living in a place that is already planned for someone else or living in a place for your own family and experience the difference in life.

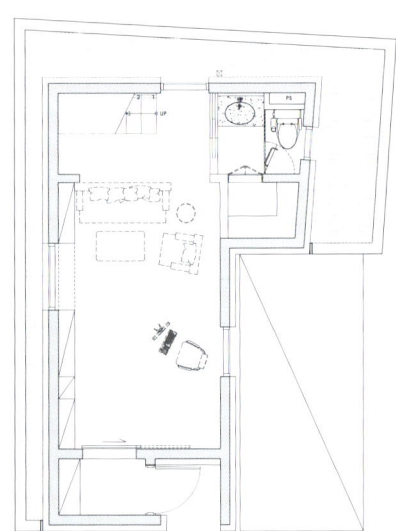

1ST FLOOR PLAN

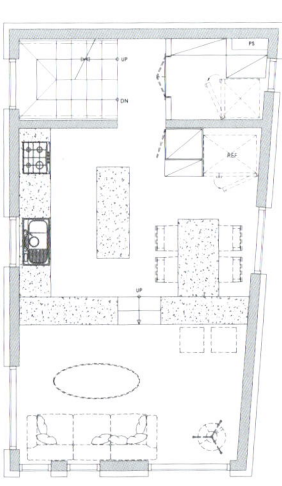

2ND FLOOR PLAN

H4912

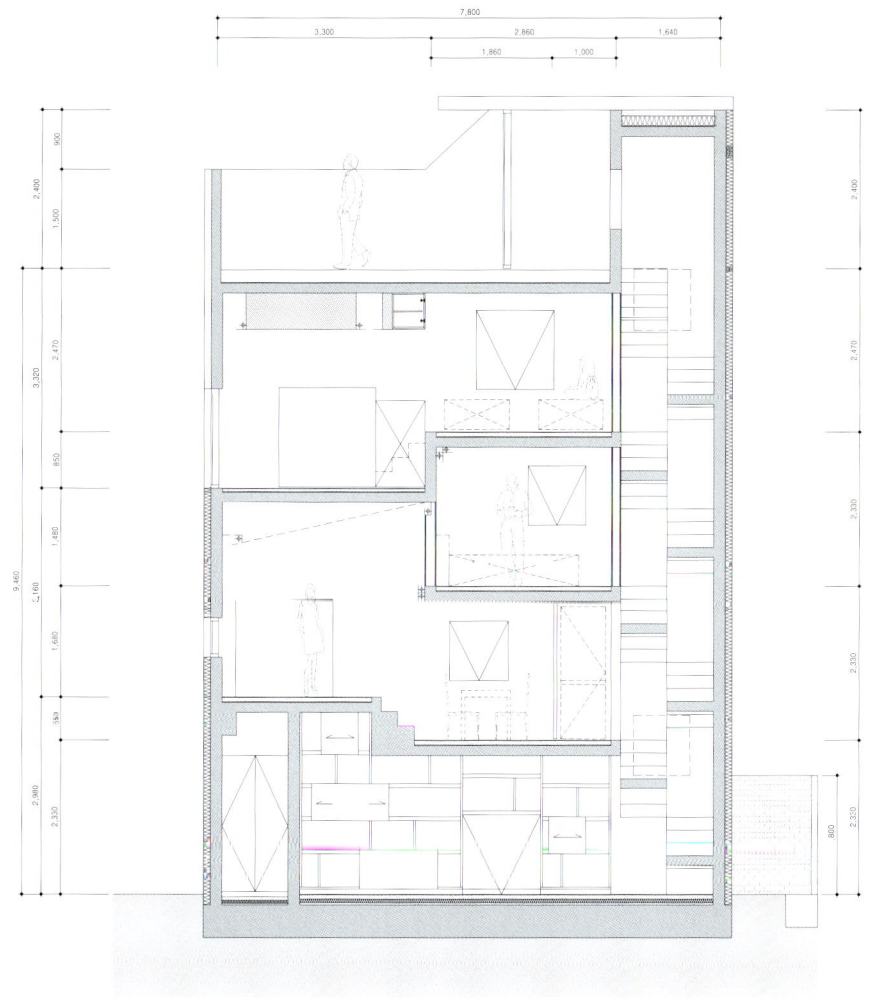

SECTION

$35m^2 - 41m^2$

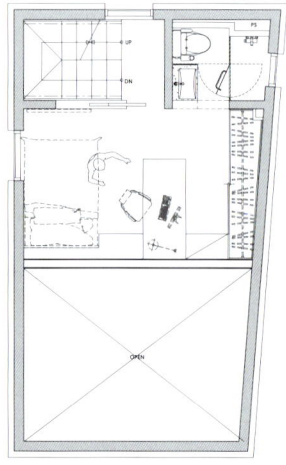

3RD FLOOR PLAN

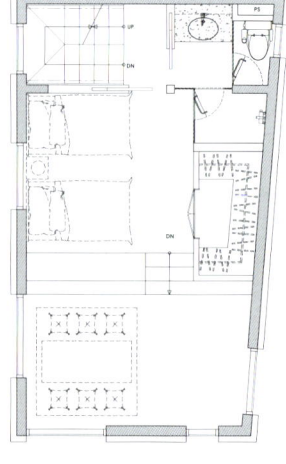

4TH FLOOR PLAN

ROOF PLAN

40m²

Big window Roof overhang

Location
Stockholm, Sweden
Use
Summer house
Site Area
2,500m²
Built Area
40m²

Exterior Finish
Wooden facade painted in traditional black Swedish Falu-paint
Interior Finish
Pine plywood boards painted brown

Project Architect
Konrad Krupinski,
Katarina Krupinska
Photographer
Åke E:son Lindman

Summerhouse T

자연을 향한 개방형 구조를 통해 다양한 공간경험을 제공하는 썸머하우스T

Krupinski/Krupinska Arkitekter

$35m^2 - 41m^2$

벽면 중 2개는 미닫이창으로 이루어져 있어서 더운 여름에 집 안으로 기분 좋은 바람이 들어오며, 돌출형 지붕이 햇볕을 가리면서 거실이 외부로 확장된다.

이 작은 주택은 스톡홀름 군도에 있는 한 호수 근처에 자리 잡고 있다. 1900년대 초반 정원사가 이곳에 살았는데, 그때 조성된 정원과 담벼락이 오늘날에도 상당 부분 남아 있다. 지난 25년 동안 정원에 관심이 많은 한 가족이 이 주택을 여름 별장으로 사용했다. 어머니와 두 딸로 이루어진 이 가족은 최근 두 딸이 새로 가정을 꾸려 가족 수가 늘어나면서 더 많은 공간이 필요했다. 자매 중 한 명은 기존의 게스트하우스를 개조하기로 했고, 다른 한 명은 어른 2명과 아이 2명이 머물 소형 주택으로 Summerhouse T를 새로 짓기로 했다. 가족은 주방, 욕실, 식사 공간, 거실, 침실 4개, 창고 공간으로 이루어진 건축 계획을 세웠다. 그런데 지역 규정상 $40m^2$을 초과하는 건물의 건축은 제한되어 있었다.

건축가가 개방된 정사각형 형태의 넓은 방과 침실, 주방, 창고를 포함하는 분리 공간을 만들고, 건물 남쪽을 따라서 식사 공간과 생활 공간을 구성했다. 지붕과 바닥, 벽은 어두운 색상으로 만들어 건물의 친근감을 높이고 시선이 주변 경관을 향하게 했다. 이 집에서 유일한 (작은) 문은

Summerhouse T

분리된 공간인 욕실로 이어진다. 건축가는 근처에 있는 바위를 고려하여 그 간격에 그대로 맞추어 욕실을 설계했다.
천장에 높은 창이 있는 개방형 구조 덕분에 놀랍도록 다양한 공간 경험을 할 수 있다. 파사드를 따라 자유롭게 이동할 수 있어서 실제보다 집이 커 보이고 넓어 보이는 느낌을 준다. 부부의 침실에서는 시골 풍경이 보이고, 안쪽으로 들어가 있는 형태로 포근한 느낌을 주는 아이들 침실에서는 할머니 집이 보인다. 집 뒤쪽의 창고로 가는 길은 한쪽이 바깥쪽 숲으로 이어져 있고 반대쪽은 평행 구조의 주방으로 이어져 대비를 이룬다. 식사 공간이 있는 주방과 라운지는 주변으로 180도 트여 있는 풍경을 선사한다. 벽면 중 2개는 미닫이창으로 이루어져 있어서 더운 여름에 집 안으로 기분 좋은 바람이 들어오며, 들쭉형 지붕이 햇볕을 가리면서 거실이 외부로 확장된다. 비가 오면 처마를 따라 빗물이 마치 커튼처럼 부드럽게 흘러 내리는데, 이때 창을 빠르게 안쪽으로 닫을 수 있다.

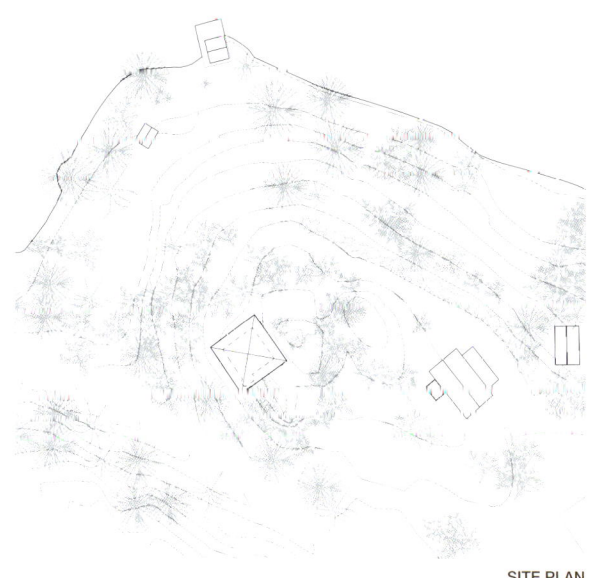

SITE PLAN

$35m^2 - 41m^2$

Two of the walls are sliding windows that on hot summer days easily opens and allow for a pleasant breeze through the house and at the same time extend the living area to the exterior, where the roof overhang provides protection against the sun.

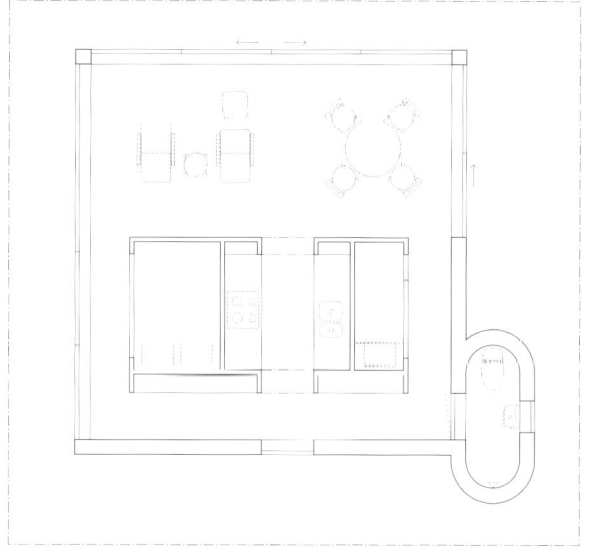

PLAN

The small house is situated by a lake in the Stockholm archipelago, on a site that in the early 1900s was inhabited by a gardener. Plants and paved walls that were then organised still remain to a large extent today. For the last 25 years another garden interested family have used the old existing house as a summer home. The family, consisting of mother and two daughters, has grown in recent years when the two daughters have formed families on their own. That in turn has made more space necessary. One sister has renovated an existing guest house, while the other sister decided to build a new small house, Summerhouse T, for her family consisting of two adults and two children. The family formulated a program that included a kitchen, bathroom, dining area, living room, four beds and room for storage. While municipal constraints did not allow a building that exceeded 40 m².

The architects created a large open square room with a standalone volume containing the beds, kitchen and storage. Along the buildings southern side is a dining area and a living area. Roof, floor and walls have a dark colour to enhance the building's intimate character and direct the eye towards the surrounding view. The house's only (trap) door leads to the bathroom, a separate volume which design was adapted to fit into an existing gap in the nearby rock.

The open organization with ceiling high windows create a surprisingly large number of different spatial experiences. Freedom of movement along the entire facade gives a feeling that the house is larger and more spacious than it in reality is. The parents' sleeping area has a view towards the countryside and the lake while the children's more enclosed sleeping alcove directs the view towards the grandmother´s house. The storage passage at the back of the house opens up to the greenery outside at one end while the parallel kitchen has contrasting views in opposite directions. The living room with dining area and lounge, in turn, have a 180 degree view towards the surrounding. Two of the walls are sliding windows that on hot summer days easily opens and allow for a pleasant breeze through the house and at the same time extend the living area to the exterior, where the roof overhang provides protection against the sun. On rainy days, water runs along the eaves almost as physically present curtain and the house suddenly closes inward.

SECTION

Krupinski/Krupinska Arkitekter

interview

큰 창을 통해 내외부의 경계가 불분명해지고 공간이 더 넓어 보이는 느낌을 준다. 확장된 지붕으로 실제보다 건물이 더 넓어 보인다.

Q. 클라이언트의 요구사항은 무엇이었으며, 그에 따른 건축적 대응은 어떤 것이었나?

A. 아이 2명을 둔 클라이언트가 아이들과 부부, 창고, 주방, 욕실이 포함된 광범위한 건축 계획을 요청하였지만 지역 규정상 40m^2를 초과하는 건물을 지을 수 없었다. 그래서 건물의 공간 내에 작은 공간을 여러 개 만드는 조밀한 구조를 제안했다.

Q. 협소한 건축대지 위, 수납공간을 창출하기 위한 아이디어는 어떤 것이었으며, 프로그램 공간구성을 어떻게 하였는가?

A. 건물 파사드를 따라 자유롭게 이동할 수 있어서 실제보다 공간이 넓게 느껴진다. 주방 공간을 통해서도 이동이 가능하므로 8자 고리 모양으로 건물에 드나들 수 있다. 막다른 길이나 우회 길이 없어 따라 걸을 수 있는 길이 길게 느껴지고, 마음대로 선택해서 드나들 수 있어 자유로운 느낌이 든다. 커다란 창문을 통해 내부와 외부의 경계가 사라져 건물이 더욱 크게 보인다. 게다가 지붕도 확장되어 있어 실제보다 건물 경계가 훨씬 더 넓어 보인다.

Q. 예산을 줄이거나, 제한된 예산 안에서 최대한의 퀄리티를 만들어내기 위한 방법은 무엇이었나?

A. 우리는 모든 프로젝트, 특히 제한된 예산으로 진행하는 프로젝트에서는 더욱 비용과 관련된 건축의 핵심 측면을 먼저 파악하려고 노력한다. 예를 들어 Summerhouse T에서는 주변 자연과의 관계가 매우 중요했다. 그래서 예산의 상당 부분을 비용이 많이 드는 창문에 배정하고, 주택의 나머지 부분에는 일반적이며 합리적인 목조 구조를 활용했다.

Summerhouse T

Q1. What were your client's requirements and architectural approach for them?
A. The client, a family with two children, required an extensive program including living area, dining area, sleeping areas for children and parents, storage, kitchen and a bath room while the municipal constraints did not allow a building bigger than 40 sq.m. We therefore proposed a very compact organisation which in turn created smaller spaces within the big space of the building.

Q. What were your ideas to utilize the space and storage and how to zone the program in such a small area?
A. The space feels larger than it is through allowing free movement along the facade. In combination with being able to move through the kitchen area one can move into a 8-shaped loop through the building. No dead ends and alternative routs gives a sensation of longer distances and an attractive freedom of choice. The big windows blurs the difference between inside and outside which further gives the building the appearance of a larger scale. Finally the extended roof gives a sense that the limits of the building are further away than in reality.

Q. Within a limited construction budget, would you be able to share your tips to make a higher quality project?
A. In all our projects, but most importantly in the ones with limited budget, we try to early identify the most important architectural aspects in relation to costs. In Summerhouse T for example the relation to the surrounding nature was very important. We therefore allowed a large portion of the budget for the expensive windows while we used a traditional and very rational wooden construction for the rest of the building.

40.7m²

Loft
Terrace
Band window

Location
Niigata-pref, Japan
Use
House
Site Area
191.0m²
Built Area
40.7m²
Total Floor Area
75.2m²

Floor
2F
Exterior Finish
Western redcedar board
Interior Finish
Chaff wall, lauan plywood

Project Architect
Takeru Shoji, Yuki Hirano
Construction
Kurita Corporation
Photographer
Isamu Murai

Y House

연창을 통해 주변 자연을 받아들인 Y 하우스

Takeru Shoji Architects. Co., Ltd.

$35m^2 - 41m^2$

Y house

거실 양쪽 끝에는 높인 상단(loft) 부분을 두어 계단참으로도 활용했다. 큼직한 연창을 통해 바람결에 살랑이는 나무를 매일 감상할 수 있고 심지어는 거실 끝 테라스에서 직접 나뭇가지를 만질 수도 있다.

신사로 가는 길목을 향하는 이번 프로젝트의 부지 바로 옆에는 200년이나 된 거대한 느티나무가 줄지어 있었다. 이 특별한 환경을 그저 바라보고 감탄하는 데 그치지 않고 그 느낌을 일상 속으로 녹여내 보고자 하는 것이 이번 프로젝트의 출발점이었다.

고객의 요청사항은 두 가지였는데 우리의 생각과 부합하는 내용들이었다. 첫 번째는, 집이 일단 단순해야 했고, 일생을 살아가면서 낡고 변해가는 모습을 즐겁게 지켜볼 수 있길 바랐다. 두 번째는 큰 느티나무와 신사로 가는 신입로가 만들어낸 이 특별한 환경을 최대한 잘 활용하는 것이었다. 느티나무의 높이가 20미터에 달했기에 일반적인 구조로 집을 짓고 평범한 창을 달면 1층에서 그 푸르름을 감상하기가 어려웠다. 그래서 거실 층고를 3.5미터로 높이고 벽의 상부에 커다란 연창을 시공했다. 거실 양쪽 끝에는 높인 상단(loft) 부분을 두어 계단참으로도 활용했다. 큼직한 연창을 통해 바람결에 살랑이는 나무를 매일 감상할 수 있고 심지어는 거실 끝 테라스에서 직접 나뭇가지를 만질 수도 있다.

여름에는 나무 사이로 스며든 햇살이 거실을 밝힌다. 가을에는 식탁 위로 색색의 낙엽이 내려앉는다. 이렇게 이 집에서는 마치 숲에서 식사하고 잠을 자고 책을 읽고 놀이를 하는 것만 같은 자연 속에서의 생활을 누릴 수 있다. 건물 외관은 미국산 삼나무로 마감했고 바닥에는 삼나무 패널을 깔았다. 층계는 연철과 나왕으로 시공했다. 아름드리 나무를 품은 이 집은 자재를 있는 그대로 사용하고 시간의 흐름에 따른 변화를 수용한다. 세월과 함께, 사람과 함께 천천히 나이 들어 가는 아름다운 집으로 남길 바란다.

SECTION B

▼ Building Maximum Height
▲ Eaves Height
▼ Second Floor Level
▼ Mezzanine Floor Level
▼ Living Floor Level
▲ First Floor Level
▲ GL
211.4
2880.8
7699.2
7487.8
1315
2292
500
150
Spare room
CH(L)=2128
CH(H)=2782
Corridor
Work Space
Living
CH=3542
Garden
1818
2727
1958
6503

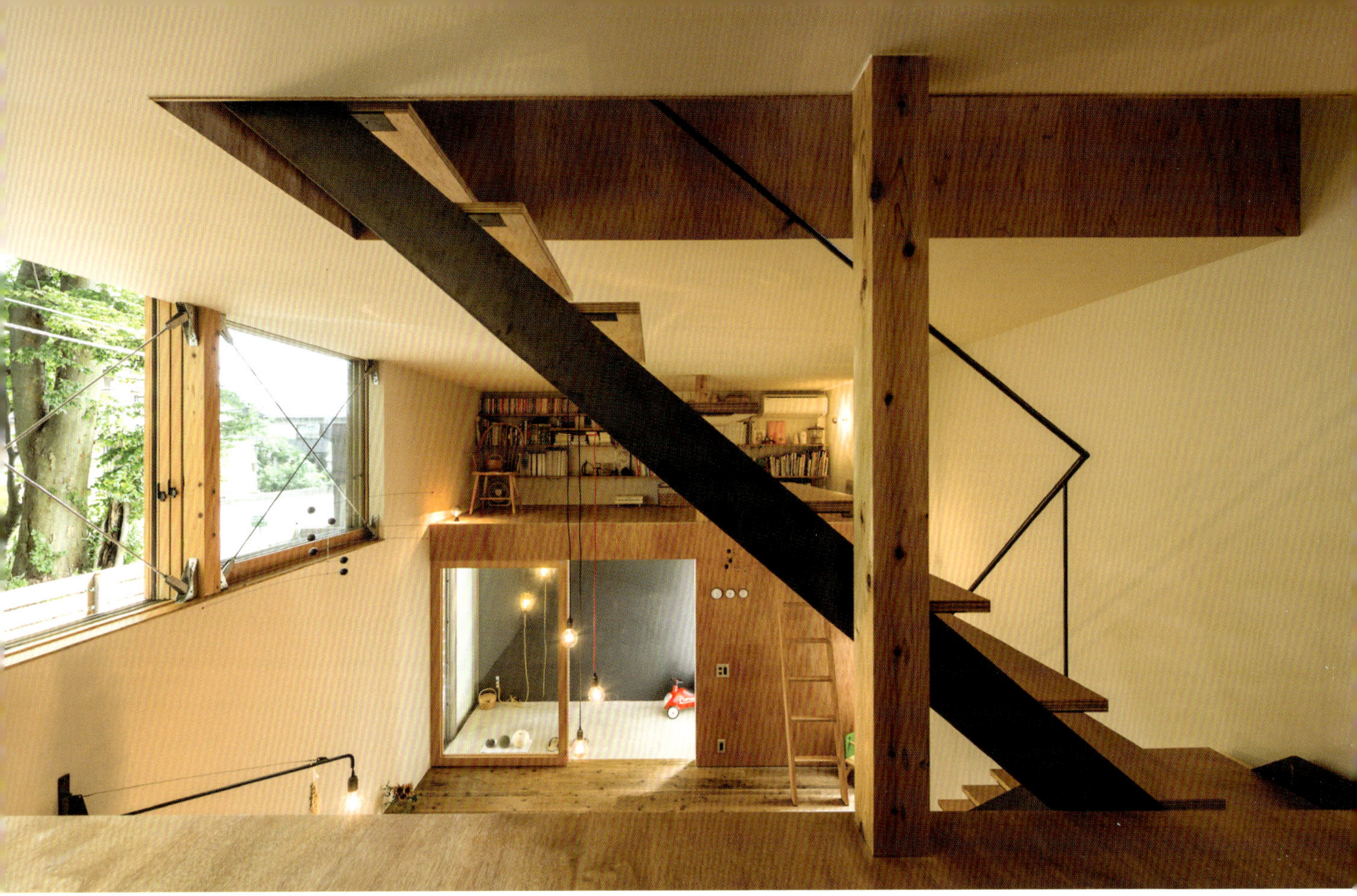

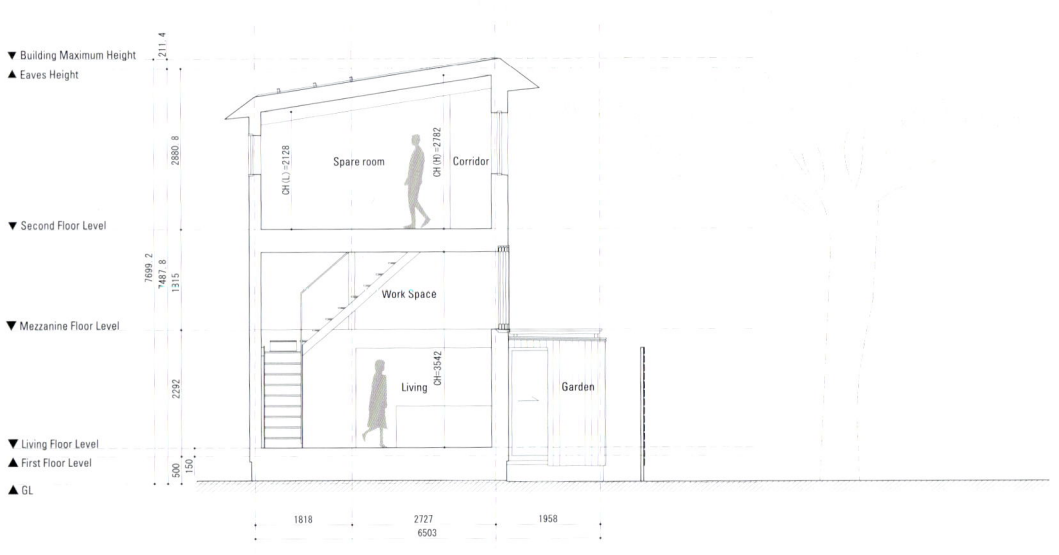

SECTION A

At the both ends of the living room, there are loft spaces used also for a landing space of the stair. We look up swinging trees in the breeze from the large band-window in everyday life, and even we can touch the trees directly from the terrace at the very end of the loft space.

Facing to the approach to a shrine, the site is found just beside gigantic 200-year-old Japanese zelkova trees. How can we bring this richness of surroundings as not only invaluable experience in the characteristic context but also richness of feelings into the daily life? That's the starting point for the project. Appropriately, the requirements from the client are two: Firstly, the house needs to be just simple and capable of any future changes with enjoyable aging process of the house in their lifetime. And the other is to make good use of the richness of the site: gigantic Japanese zelkova trees and the approach to the shrine. Since the Japanese zelkova trees are 20 meters in height, in a conventional way, we cannot get a view of verdant branches from mundane windows on the ground floor. Therefore living room has 3.5 meters of ceiling height, and moreover a large high-side band-window.

At the both ends of the living room, there are loft spaces used also for a landing space of the stair. We look up swinging trees in the breeze from the large band-window in everyday life, and even we can touch the trees directly from the terrace at the very end of the loft space. Sunlight filtering through the trees enters in the living room in summer. Colorful leaves fall on the table in autumn. Consequently, the house is obtained the environment and experience as if we ate, slept, read and played in the forest. The exterior is clad in western red cedar, the floor is covered with scaffolding cedar boards, and the stair is made with wrought iron works and lauan. Using the materials as it is, accepting the changes as time goes by, House to catch the tree is aimed at being an ever-changing house with the life style of the clients and the seasons.

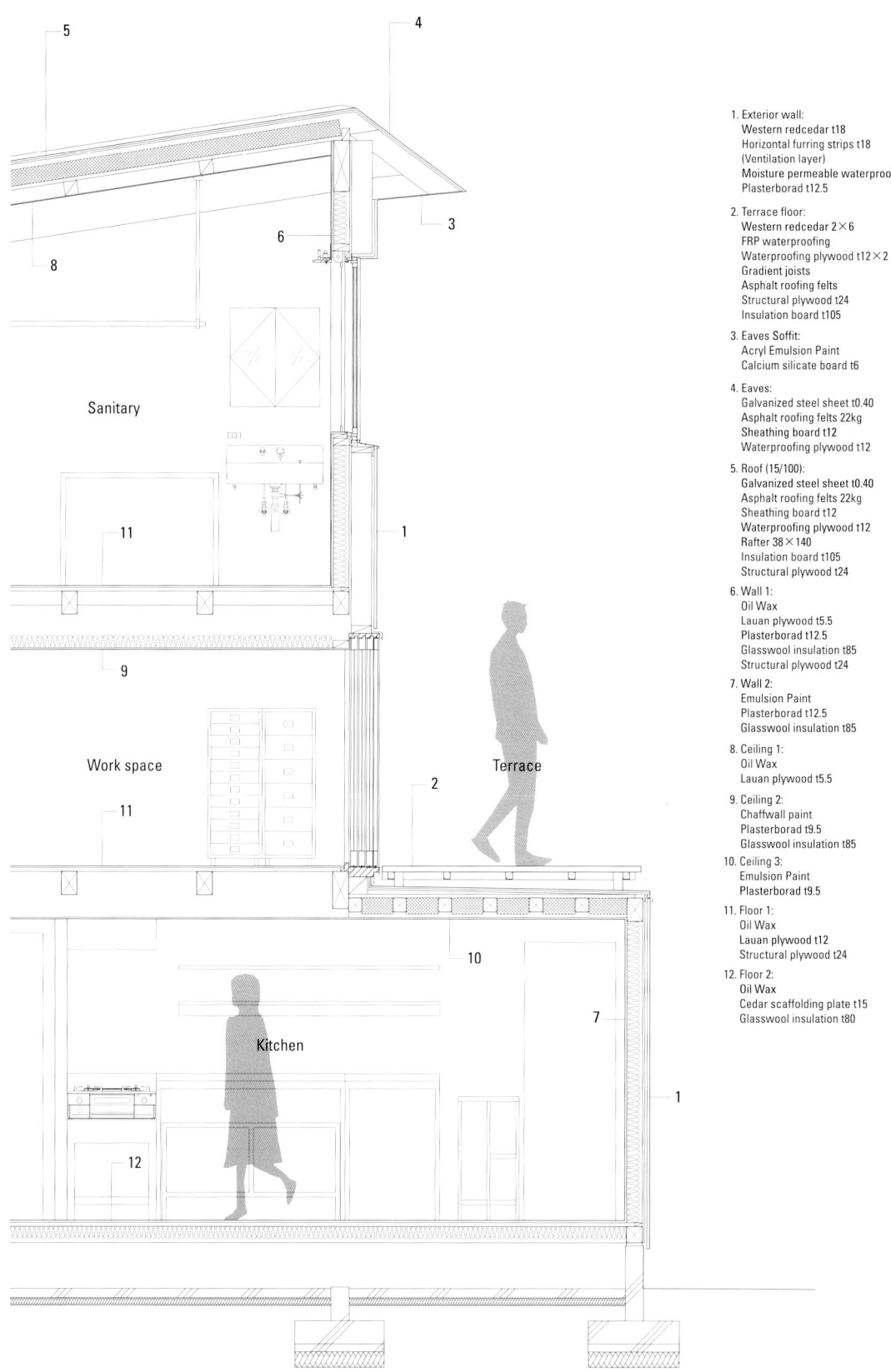

PARTIAL SECTION DETAIL

$35m^2 - 41m^2$

Y house

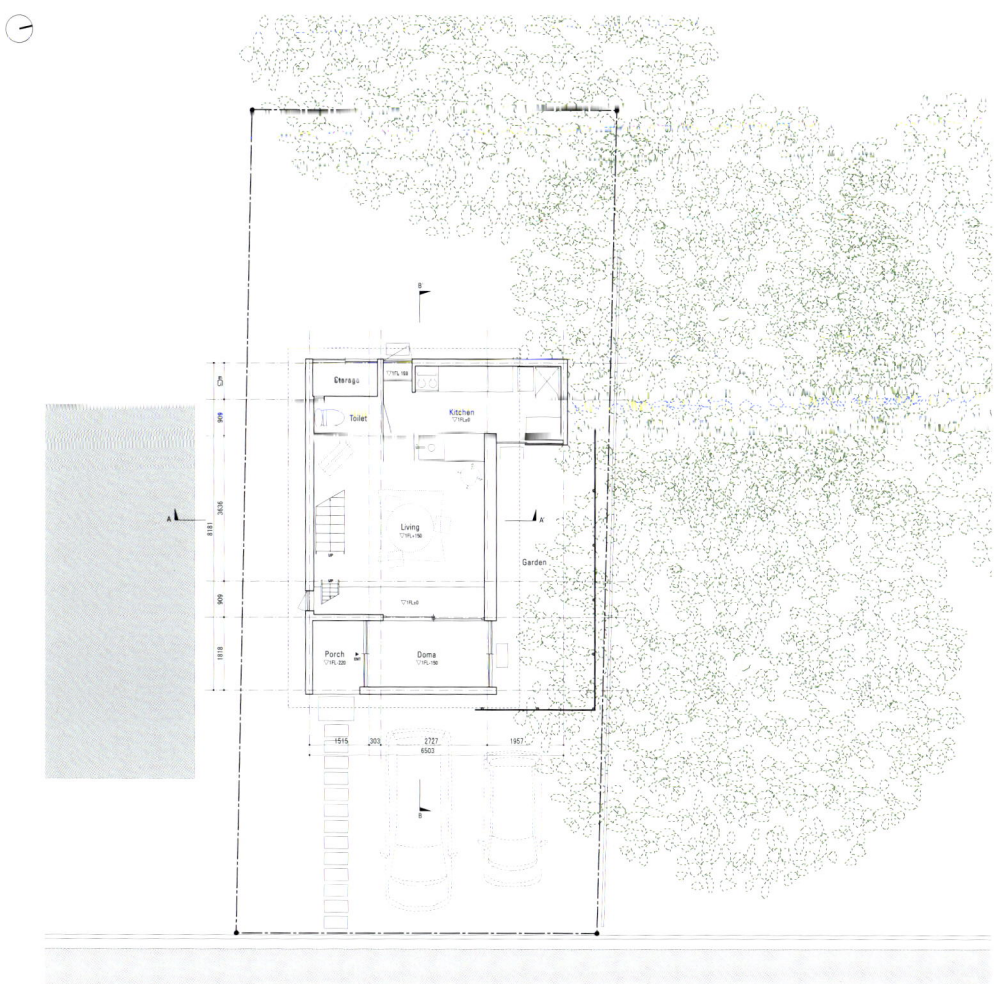

1ST FLOOR PLAN

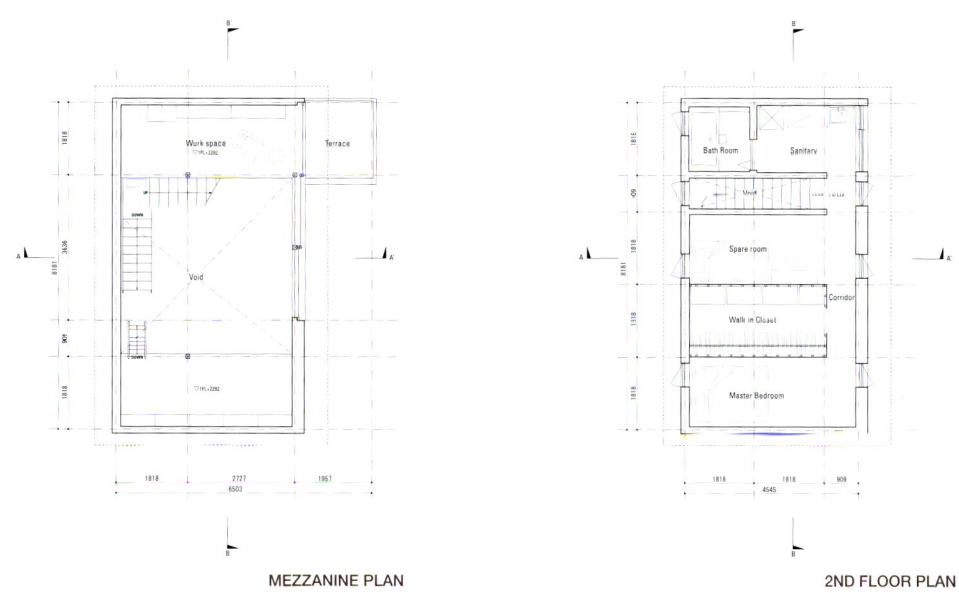

MEZZANINE PLAN

2ND FLOOR PLAN

39.26m²

Urban-regeneration Narrow

Location
Busan, South Korea
Use
Neighbourhood living facility, accommodation facilities
Site Area
53m²
Built Area
30.26m²
Total Floor Area
133.08m²

Floor
4F
Structure
Exposed concrete
Exterior Finish
Reinforced concrete
Interior Finish
V.P on Plaster Board

Project Architect
Oh Sin-wook
Construction
Taebaek Construction
Photographer
Yoon Joonhwan

Vicolo

공생의 건축을 담은 협소건물 비꼴로

RAUM ARCHITECTS GROUP (라움건축)

$35m^2 - 41m^2$

Vicolo

골목의 공간에 건축공간을 덧대어서 골목계단의 감성이 건축 공간으로 확장되게 하였다. 그러므로 비꼴로의 시선은 골목을 보게 하였고, 동선은 골목길을 이용하여야 한다.

오랜 골목길을 되살리고, 이 길과 협소빌딩의 내부공간을 연결하였다. 이 공간, 이 건축물을 Vicolo(뒷길, 골목길, 샛길)로 부른다.
비꼴로의 디자인 목적은 오래된 골목계단의 재생이며, 골목의 경험이 더 풍성해지도록 하는 것이었다. 그래서 골목의 공간에 건축공간을 덧대어서 골목계단의 감성이 건축 공간으로 확장되게 하였다. 그러므로 비꼴로의 시선은 골목을 보게 하였고, 동선은 골목길을 이용하여야 한다.
노출콘크리트는 화려한 시공기술을 바탕으로 하는 물성의 표현이다. 하지만 이를 시도하기에는 많은 노력과 치장이 수반되어야 한다. 일반적으로 콘크리트 구조체를 위한 거푸집공법을 그대로 노출시키는 것은 새로운 대안이 될 수 있다. 이에 초량 비꼴로는 적정한 공법, 골목에 적합한 이미지, 적절한 공간의 크기, 일체화된 연속성과 스케일 등을 찾아내면서 완성되었다. 낮은 층고, 휴먼스케일, 주변의 작은 매스와 볼륨, 색상, 오랜 빛깔 등은 그 완성도를 높여 준다.

또한 비꼴로는 절제된 볼륨에 가해진 최소한의 다듬기를 통해 멀리서 보면 화려하지만, 가까이 다가가면 소박한 재료와 이미지가 표현되도록 하였다. 가까이 다가가며 진정성이 느껴지는 공간, 재료, 분위기, 규모를 최대한 찾아내는 작업이었다.

SITE PLAN

$35m^2 - 41m^2$

I extended the sentiment of an alley stairway into the building space by placing the building space in the alley space. In that way, Vicolo was put facing the alley, and now the circulation should use the alley.

I revived an old alley and connected the alley to the internal space of a small building. This space or this building is called "Vicolo" (back street, alley, or side road).

The design purpose of Vicolo was to regenerate an old alley stairway and to make the experience of an alley richer. So, I extended the sentiment of an alley stairway into the building space by placing the building space in the alley space. In that way, Vicolo was put facing the alley, and now the circulation should use the alley.

Exposed concrete is an expression of property based on a splendid construction technique. However, it should be accompanied by lots of efforts and decorations to try this.

Generally, the method of exposing a form for a concrete structure can be a new alternative. As such, Vicolo was completed while an appropriate technique, image that suits the alley, proper space size, integrated continuity, and scale were found. Low story height, human scale, neighboring small masses and volume, colors and old hues heighten the completeness.

Also, Vicolo looks very showy from a distance, but humble materials and images are expressed when you get up close. This could be obtained with minimum trims applied to a restrained volume. I worked as much as I could to find the space, materials, atmosphere, and size from which you can feel sincerity when you get up close.

$35m^2 - 41m^2$

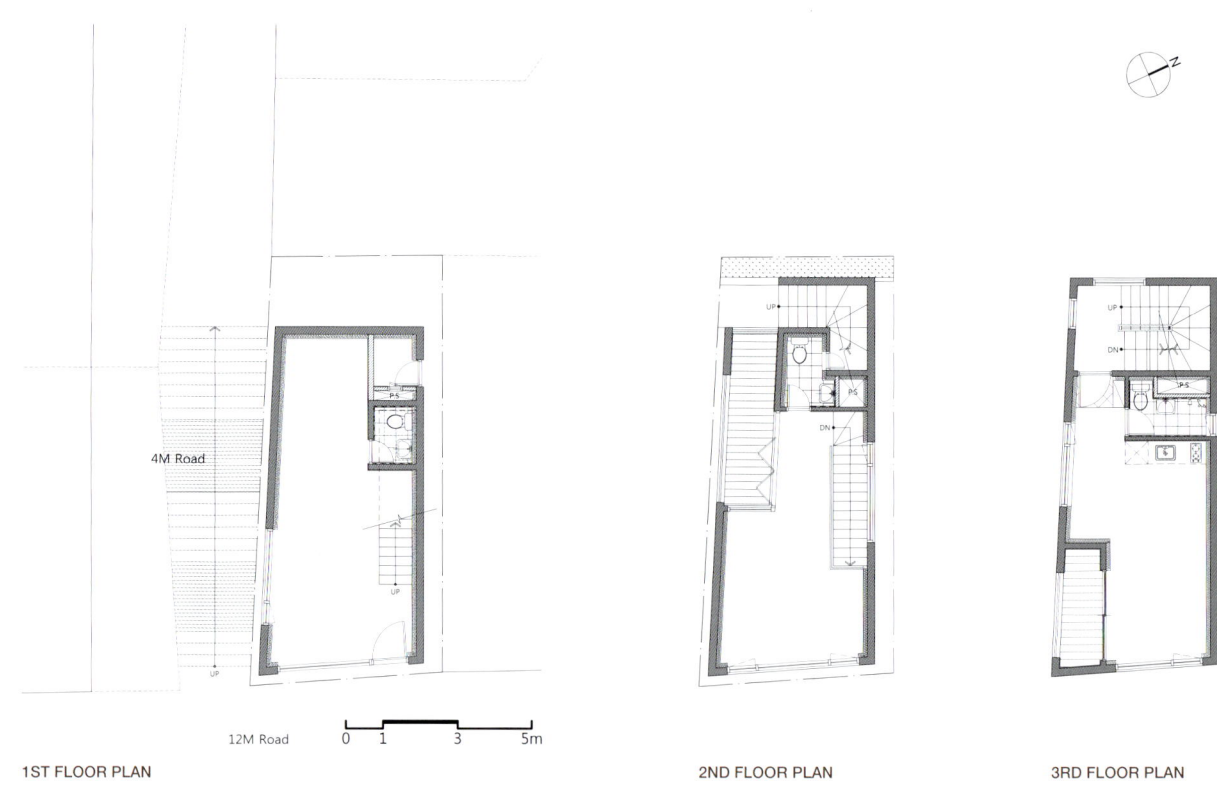

1ST FLOOR PLAN	2ND FLOOR PLAN	3RD FLOOR PLAN

Vicolo

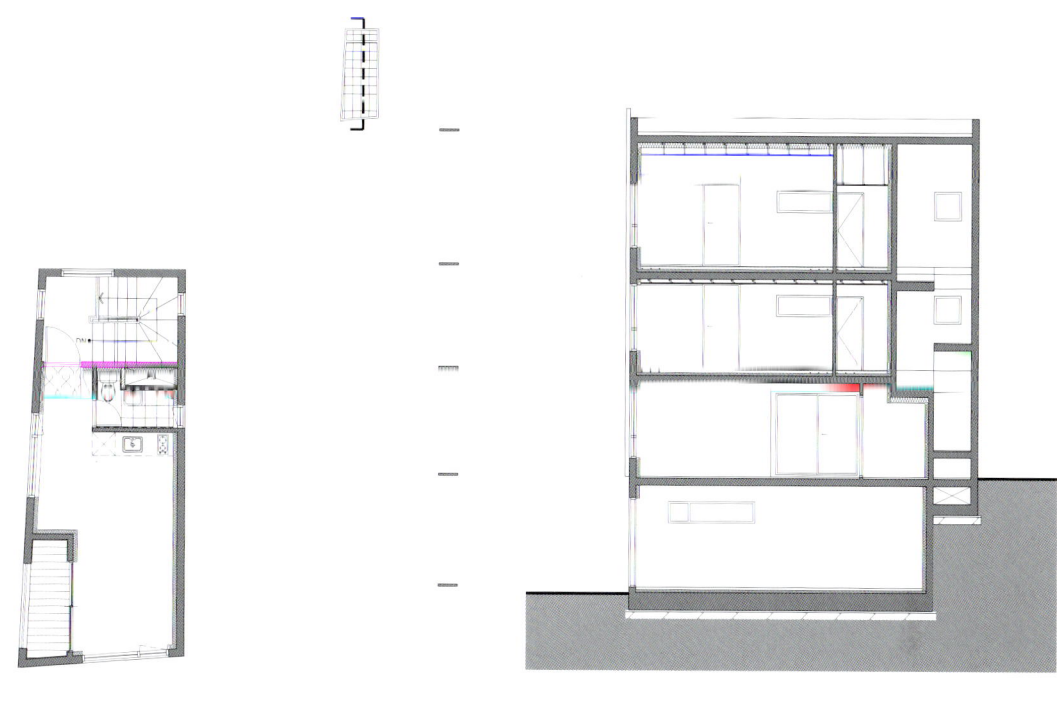

4TH FLOOR PLAN

LONGITUDINAL SECTION

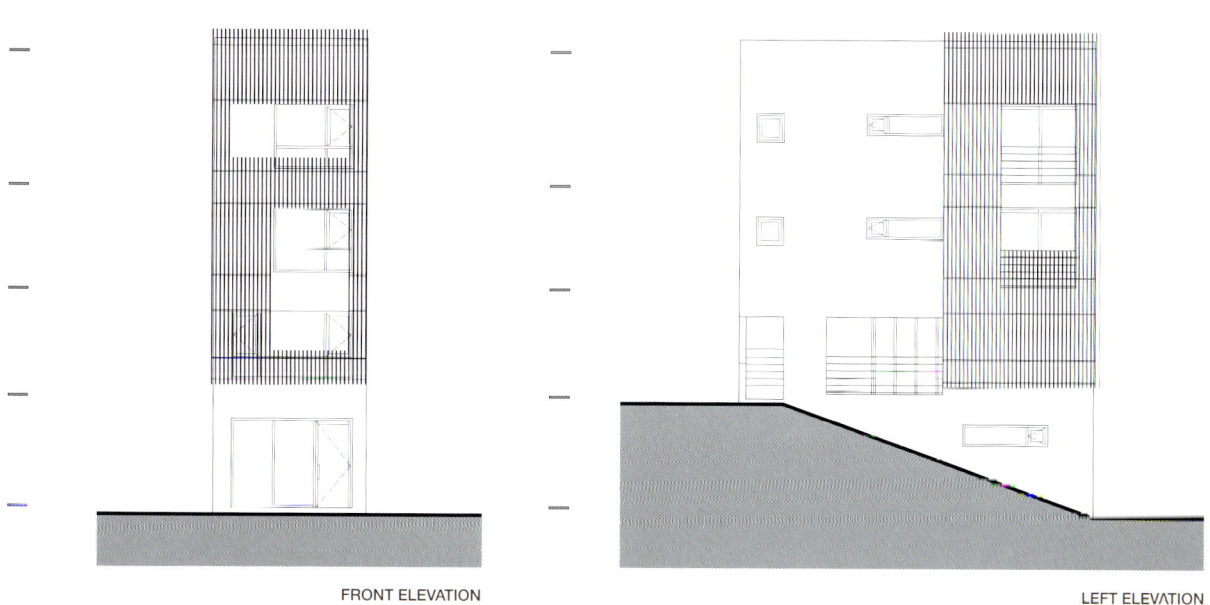

FRONT ELEVATION

LEFT ELEVATION

$35m^2 - 41m^2$

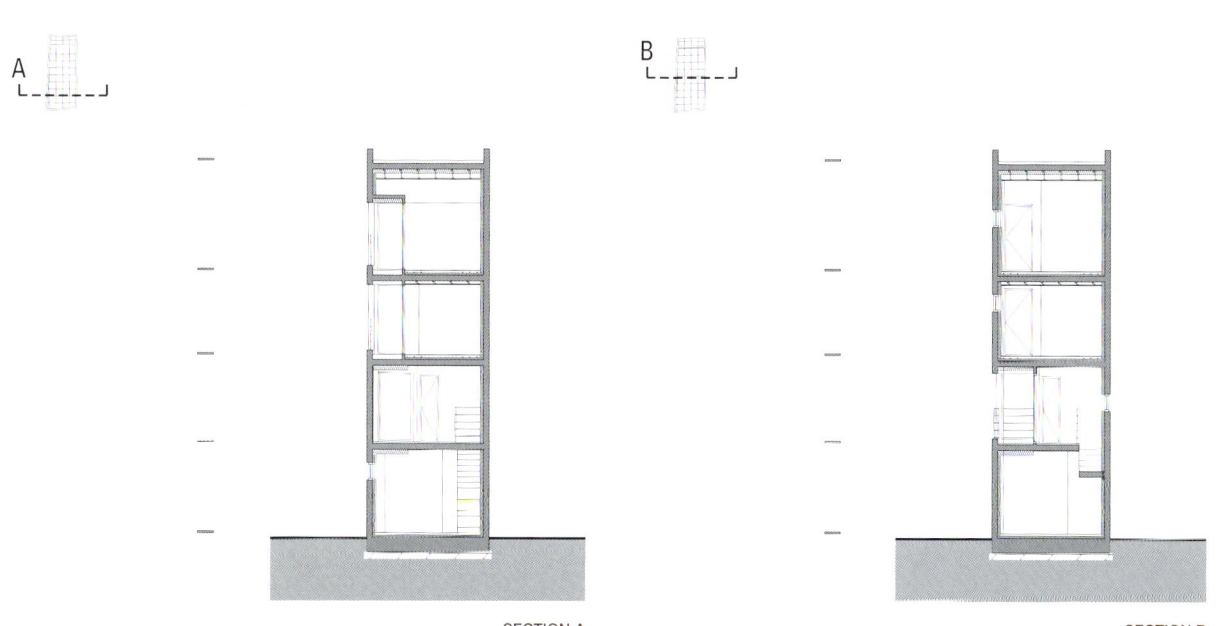

SECTION A

SECTION B

RAUM, Oh Sinwook

interview

**협소주택은 분리된 작은 공간들이 수직적으로 연결되거나,
새로운 라이프 스타일을 이끌어내는 아이디어로 만들어진다.
또한 하나의 공간적 성격만을 가지는 것이 아니라,
좁아도 다양한 기능을 동시에할 수 있도록
되어있고, 기능별·층별 공간적 겹침이 있어서
강한 연속성을 가진다.**

Q. 협소주택에 대한 정의를 어떻게 생각하고 있는지 궁금하다.
A. 일반적으로 공간의 단위는 계획 각론적으로 적절한 면적의 크기와 높이 등이 있다. 그리고 주거는 기능에 따라 규정되는 연결성이 있다. 그래서 주거의 공간은 최소한의 기능연관성들이 공간적인 연속성을 가지면서 구성되어 있다. 그러나 대지의 상황이 가진 특수성(자투리, 틈새, 비정형 등)으로 적절한 각론적 규모의 공간을 만들기 힘든 경우가 있고, 한 층에 가용공간이 좁아서, 기능적 연관성이 있는 공간을 수평적으로 연속 구성하기 어려운 상황이 있다. 이때 주어진 규모의 공간과 형태 안에서 새로운 방식으로 공간을 구획하고, 연결하면서 만들어진 집이 협소주택이다. 특히 협소주택은 분리된 작은 공간들이 수직적으로 연결되거나, 새로운 라이프 스타일을 이끌어내는 아이디어로 만들어진다. 또한 하나의 공간적 성격만을 가지는 것이 아니라, 좁아도 다양한 기능을 동시에 할 수 있도록 되어있고, 기능별, 층별 공간적 겹침이 있어서 강한 연속성을 가진다. 그래서 협소주택은 좁은 땅이지만, 상황이 만들어내는 공간의 특성에 기능이 잘 베여있어서 넓은 주거공간이 가진 다양한 경험을 모두 누릴 수 있는 집이라 생각된다.

Q. 클라이언트의 요구사항은 무엇이었으며, 이에 어떻게 대응하였는가?
A. 상업지의 협소한 최소한의 주거와 예술 작업을 동시에 할 수 있는 작업실, 그리고 문화를 조성할 수 있는 대안공간을 조성하는 것이었다. 그래서 우선적으로 상업지의 좁고 긴 땅에 주차대수를

Vicolo

확보한다는 것은 공간의 효율과 경제적 가치를 떨어뜨리는 시형이 되어, 부널주치장을 설치하지 않아도 되는 규모의 건축을 제안하였다. 그래서 봉노는 1,2층은 근린생활시설, 3,4층은 생활형 숙박시설로 허가를 받았다. 생활형 숙박시설은 장기체류가 가능한 숙박시설이며, 합법적으로 게스트하우스로 운영이 가능한 시설이다. 그리고 지형(레벨차이)을 최대한 이용하여, 1,2층의 진입과 3,4층의 진입구를 분리하여, 높이 오르는 계단의 부담을 줄였다. 협소주택에서 계단의 위치와 구조는 내부공간을 규정하고, 활용하는 면에서는 상당히 중요한 요소가 된다. 그래서 계단의 구조가 공간을 분리하지 않고, 연결하는 기능을 할 수 있도록 설계하였다

Q. 협소한 대지 내에서 최대한의 공간 창출을 위한 아이디어는 무엇인가?

A. 협소주택은 주변의 환경(땅의 특성)을 최대한 이용하여, 건축공간과의 관계를 맺어주어야 내부공간이 극대화 된다. 그래서, 창의 위치, 지형의 이용, 계단의 위치 등을 심각하게 고민하면서 설계하였다. 특히 비꼴로는 솨속의 골목과 오랜 계단을 건축의 내부공간으로 끌어오고, 내부에서 그 공간을 시선적으로 응시할 수 있도록 하여, 좁은 공간의 한계를 극복하였다. 특히 2층에 만들어진 발코니는 골목을 향해 과도할 만큼 크게 만들어졌고, 이는 내부공간과 외부공간의 접점공간으로 다양한 공간적 확장효과를 만들어낸다. 내부에서 최대한 채광을 유지하도록 하였고, 높은 층고와 더불어, 도시의 시선이 확보되는 곳에 창을 만들어, 시선의 확상을 시도하였다. 협소수백입수록 실제 사용하는 수방과, 욕실의 공간은 적정크기를 유지하여, 오히려 생활히면시, 기능적으로 풀썬아수나, 심리석으로 삭은십이라는 느낌를 빙어일 수 있노록 하였다.

Q. 수직적인 생활에서 오는 사용자의 불편함을 극복하기 위한 대안은 무엇인가?

A. 오르내림의 불편함은 적당한 움직임보다는 연속된 기능이 단절될 때 더 큰 불편함이 생긴다. 그래서 반드시 이어져야 하는 기능은 같은 레벨에서 이어주고, 적당한 거리를 두어도 가능한 기능은 수직적으로 분리하는 방식이 효율적이다. 그리고 계단의 구조는 공간을 단절시키기도 하지만, 오히려 다양한 가능성으로 공간을 연결해준다. 계단을 이용한 공간의 연결이 자연스럽고, 불편하지 않으려면, 계단의 레벨을 적절히 조절하여, 부담스럽지 않는 높이를 연속적으로 만들어주는 것이 필요하다.(스킵플로어; 다양한 레벨의 바닥연결 구조) 특히 4층이라는 공간까지의 접근은 매우 불편하다. 그래서 비꼴로에서는 지형을 이용하고, 최상층의 접근에 할애하는 계단은 두 개 층의 높이로 설계하였고, 계단의 공간에서 감성적으로 편안함을 느낄 수 있도록 창을 디자인하였다. 1,2층의 내부계단도, 기능적 한계를 넘어 오브제로, 골목길의 체험을 가능한 구조로 하여, 오히려 재미를 더하였다. 잠시 오르는 피로감을 잊을 수 있도록 하였다.

RAUM, Oh Sinwook

interview

Q. How do you define the small house?
A. Generally, the unit of space consists of an appropriate area size and height in planning details. And a residence has a connectivity defined according to functions. Therefore, a residential space is composed in a way that a minimum functional association has a spatial continuity. However, the special characteristics of situations of a site sometimes make it difficult to obtain the appropriate detailed scale of spaces, and there is a situation where it is difficult to compose a space horizontally in continuity with a functional association. Here, a house that is built by making compartments and connecting the space with a new method within a given space and form is a small house. In particular, a small house is made by connecting separated small spaces vertically or with an idea of drawing a new life style. Also, instead of having only one spatial characteristic, it has various functions despite being narrow, and strong continuity thanks to spatial overlapping by function and story. Therefore, although a small house was built on a small area of land, it can provide various experiences that a large area has as the characteristics of its spaces generated according to situations hold functions.

Q. What were your client's requirements and your architectural approach for them?
A. The task was to create a small studio that functioned as both a minimal residential space and workroom and an

alternative space that could create culture in a business area. So, since securing parking space on a narrow and long piece of land in a business area lowered the efficiency and economic value of the space, an economical building was suggested where an attached parking lot did not need to be installed. So, the first and second floors were permitted as a neighborhood living facility and the third and fourth floors as living accommodations. The living accommodations are a facility where long stays are possible and the operation of a guest house is legally possible. The burden of a staircase was reduced by separating the entrance on the first and second floors and that on the third and fourth floors, making the most of the topography (difference in levels). The location and structure of a staircase in a small house becomes an important factor in defining and utilizing the internal space. So, I designed the structure of the staircase not to be separated, but functionally connected.

Q. What were your ideas to utilize the space on small site and how to zone the program?
A. As for small houses, the surrounding environment (characteristics of land) should be used as much as possible to have a relation with the building space for the maximization of internal space. So, I thought a lot about the location of windows and the staircase and use of topography during design. In particular, the limit of small space was overcome in Vicolo by drawing the alley and old stairway on the left side into the internal space in such a way that users can look at the space from inside. In particular, the balcony created on the second floor was excessively big toward the alley. As a contact space between internal and outer spaces, this balcony makes various spatial expansion effects. The sunlight inside can be maintained as much as possible, and windows were installed in a place where the urban landscape is secured. This enables an extension of sight along with high story height. An appropriately large size was maintained for the kitchen and bathroom to avoid functional discomfort or a psychological feeling of being in a small house while living there.

Q. What are the alternatives to overcome user discomfort from vertical life in the building?
A. Walking up and down rather than appropriate movement creates bigger discomfort when the continued function is severed. Therefore, the functions that must be continued were put on the same level, while the functions could be separated, even though they have a moderate distance, were efficiently separated vertically. The continuity of space using a staircase can be natural and comfortable when the level of the staircase is properly adjusted and a comfortable height is continuously provided. (Skip floor: A structure with floor continuity on various levels). In particular, the approach to the fourth floor space is very uncomfortable. That is why topography is used in Vicolo. The staircase used to approach the highest floor was designed with two levels of height and a window was added in the staircase space for emotional comfort. In the case of internal space for the first and second floors, it was created as a structure that enabled people to have some fun with the experience of the alley as an objet beyond the functional limit. The fatigue of going up can be forgotten for a while.

40.99m²

Loft Terrace

Location
Tokyo, Japan
Use
House
Site Area
152.20m²
Built Area
40.99m²(1F) + 18.83m²(Loft), 48.75m²(2F) + 20.35m²(Loft)
Total Floor Area
128.94m²

Floor
2F
Exterior Finish
Fiber reinforced cement siding
Interior Finish
Plaster, structural plywood, white transparent color

Project Architect
Takuya Tsuchida, Kano Hirano
Construction
Kosho Kensetsu Ltd.
Photographer
Ryoma Suzuki

SUBAKO

테라스가 달린 집 수바코

no.555

35m² - 41m²

개방면 바깥쪽의 테라스는 벽으로 둘러싸인 형태로 만들었다. 남쪽 테라스는 외부의 시선에서 사생활을 보호하고 북쪽 테라스는 북풍을 어느 정도 막을 수 있도록 했다.
집 내부에는 두 군데 중간층을 두어 각종 장비나 물품 등을 보관할 수 있는 넉넉한 공간을 마련해 생활 공간이 수납을 위한 공간으로 침해되지 않도록 했다.

본 프로젝트의 부지는 도쿄시 오타구의 한 언덕 위에 위치해 있다. 전면도로와 접한 남쪽은 공간이 비교적 넉넉하고 북쪽에는 연일 아이들의 웃음소리가 끊이지 않는 공원이 있다. 멀리는 도쿄 타워와 시나가와의 고층건물들이 보이고 동쪽과 서쪽에는 이웃 주택이 인접해 있다. 우리는 부지의 각 측면을 최대한 자연스럽게 기획했다. 동·서 양측 면이 모두 이웃집 건물들로 막혀 있으므로 남측과 북측을 열린 구조로 만들었다. 남측과 북측 면을 개방된 구조로 설계하는 경우 보통은 남쪽과 북쪽의 공간 성격이 달라 그 개방 형태도 다르게 하지만, 이번에는 양쪽 모두 조건이 잘 갖추어져 있어 같은 방식으로 만들었다.
원활한 채광을 위해 넓게 개방한 남측과 마찬가지로 북측도 역시 개방면을 넓게 두어 시원한 전경을 마주하게 했다.
개방면 바깥쪽의 테라스는 벽으로 둘러싸인 형태로 만들었다. 남쪽 테라스는 외부의 시선에서 사생활을 보호하고 북쪽 테라스는 북풍을 어느 정도 막을 수 있도록 했다. 집 내부에는 두 군데 중간층을 두어 각종 장비나 물품 등을 보관할 수 있는 넉넉한 공간을 마련해 생활 공간이 수납을 위한 공간으로 침해되지 않도록 했다.
보통은 각 방의 위치와 동선 등을 고려하면 집 전체 구조가 대칭을 이루게 되기는 어려우나, 이 주택은 대칭적인 주변 환경을 고려하여 건물 자체도 대칭적인 구조를 갖도록 설계했다. 각 방의 기능은 다르지만 그렇다고 해서 중요도에 차이를 두지는 않았다. 세심하게 마련된 공간에 깔끔하게 정리된 물품들과 마찬가지로 주거자의 생활도 녹음에 둘러싸인 둥지처럼 단정하게 정돈될 수 있는 공간을 만들고자 했다.

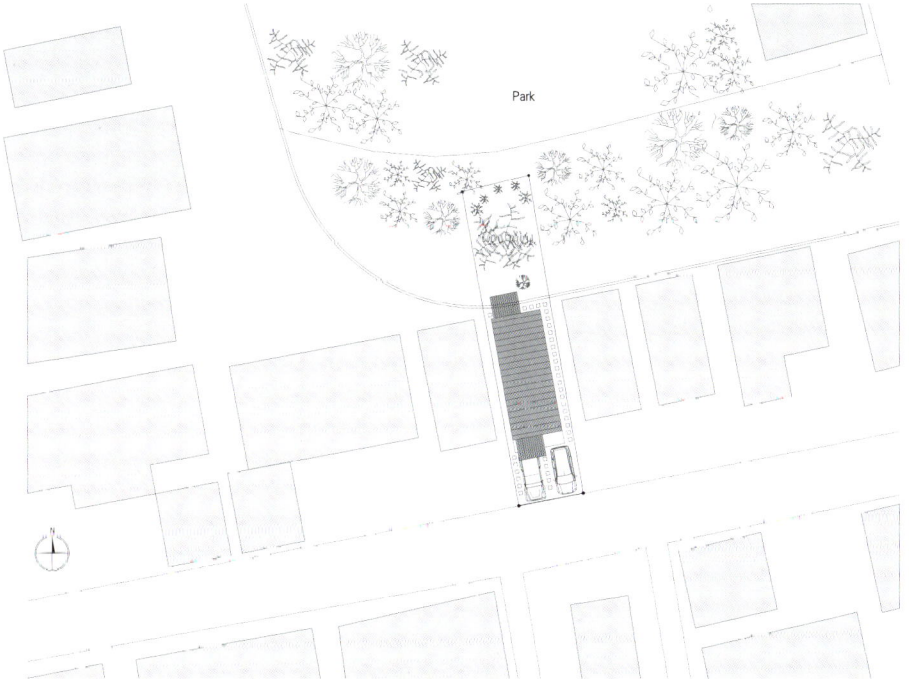

SITE PLAN

$35m^2 - 41m^2$

> We build a terrace surrounded by walls outside of the opening. It will work to avoid people's looks from outside on the south side, and to reduce the north wind on the north side.
>
> The house has two intermediate floors, giving sufficient spaces for an owner to store a number of items for work and hobby. It secures large space so that their daily life is not distracted.

The site is located on a hill in Ota ward, Tokyo.
There is a frontal road on the south side, giving enough space to the site. On the north side, you find a park filled with children's laughter every day. You can also see Tokyo Tower and a group of buildings in Shinagawa area in the distance. There are houses closely standing on the east and the west side.

We arrange each side of the site in most natural ways. While the east and the west sides are closed because of the neighboring houses, we make the south and north sides open. How are we going to develop the south and the north sides? There is usually difference in character between the south and the north, so the way to open will be changed. However, because both are well-conditioned in this case, we decide to treat them equally. We leave the south side wide-open for getting sunshine effectively and did the same for the north side for the scenery.

We build a terrace surrounded by walls outside of the opening. It will work to avoid people's looks from outside on the south side, and to reduce the north wind on the north side.
The house has two intermediate floors, giving sufficient spaces for an owner to store a number of items for work and hobby. It secures large space so that their daily life is not distracted.

Normally, due to the relationships between directions and each room, a whole house cannot be symmetric. However, as we treat the surrounding environment symmetrically, the shape of the house itself becomes symmetric. While each room plays different role, it doesn't give a feeling of priority in a good sense.
Belongings of the owner are kept with care and the life of the owner is going to continue. Just like a bird nest surrounded by green…

35m² - 41m²

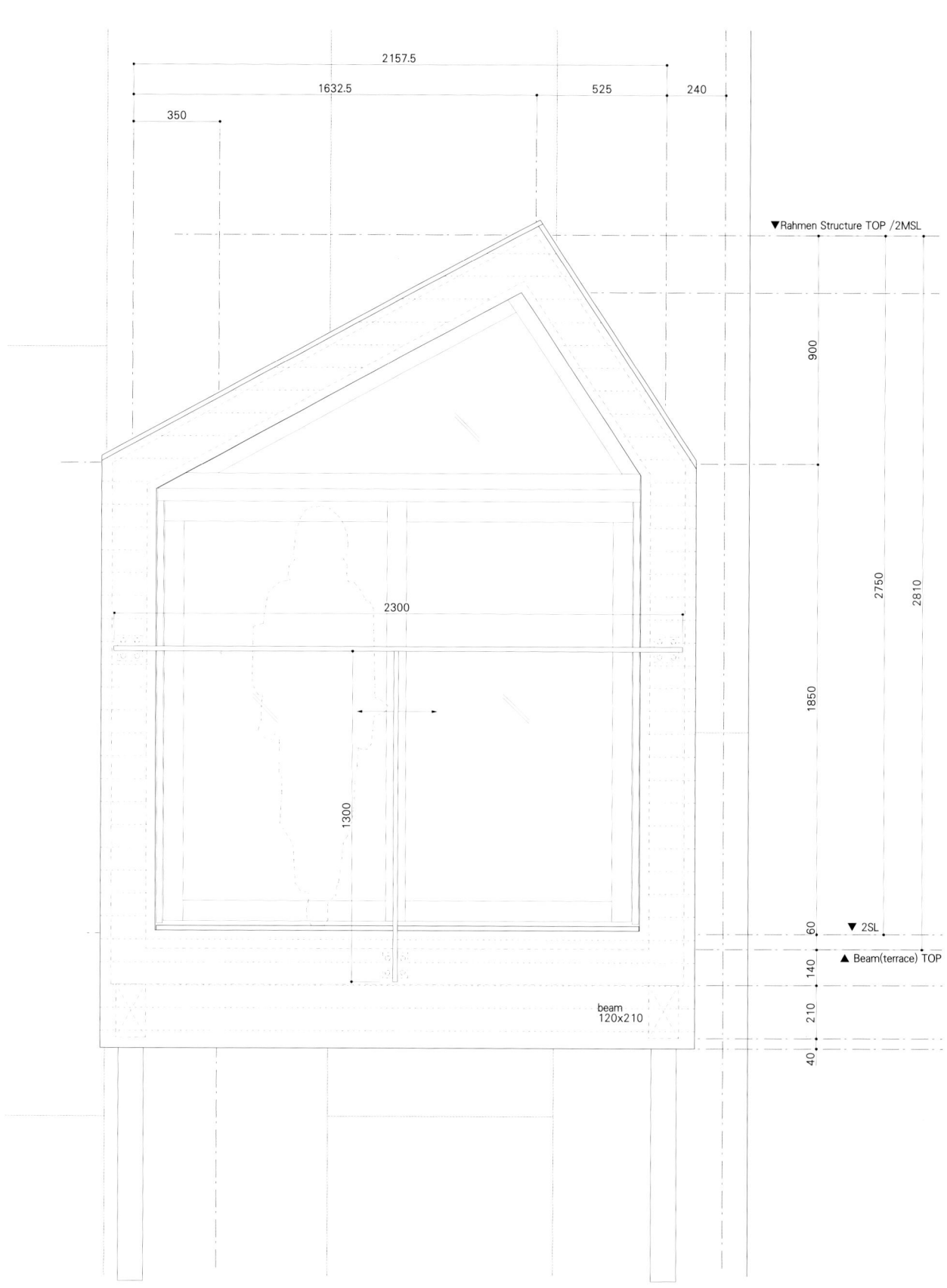

Subako

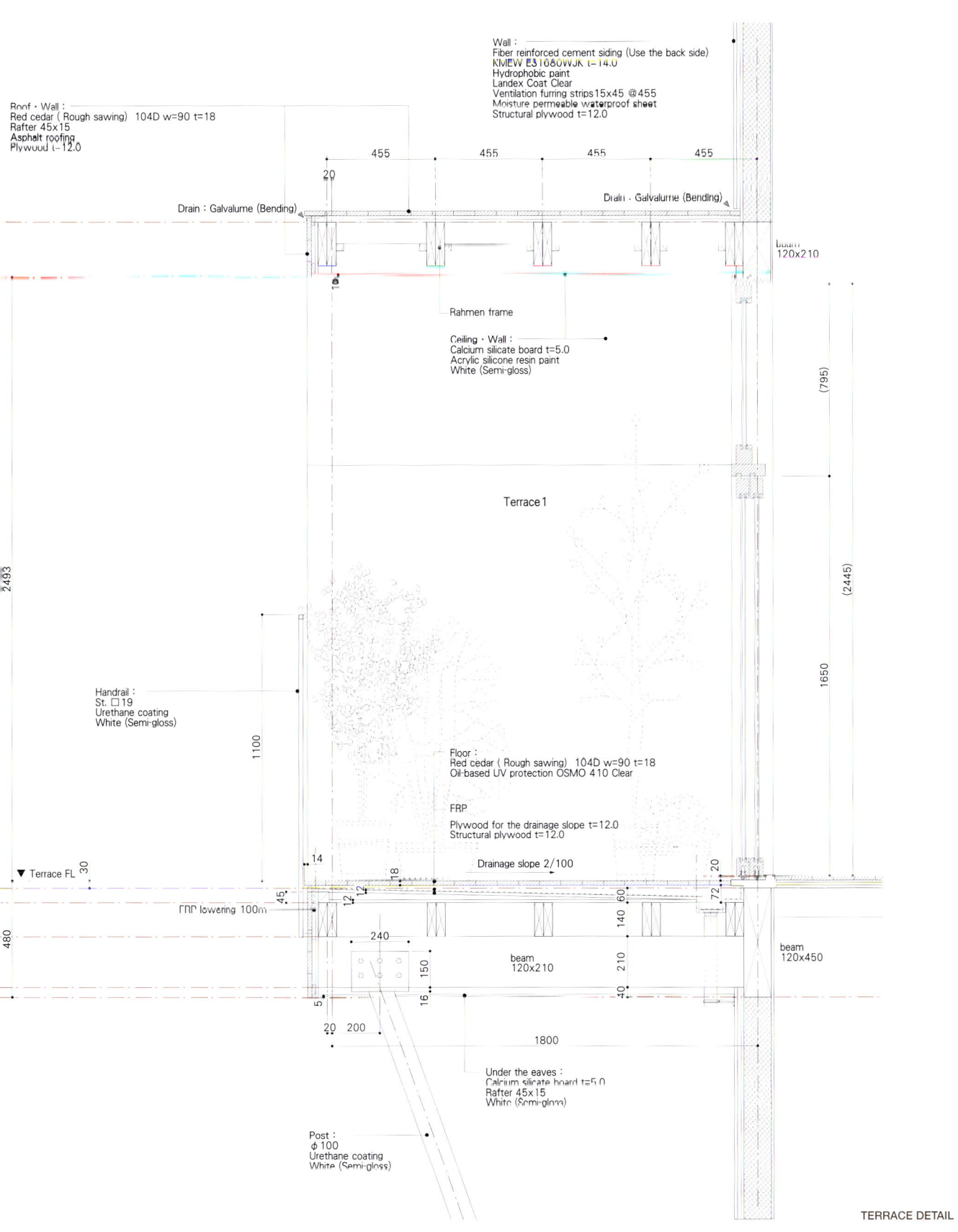

TERRACE DETAIL

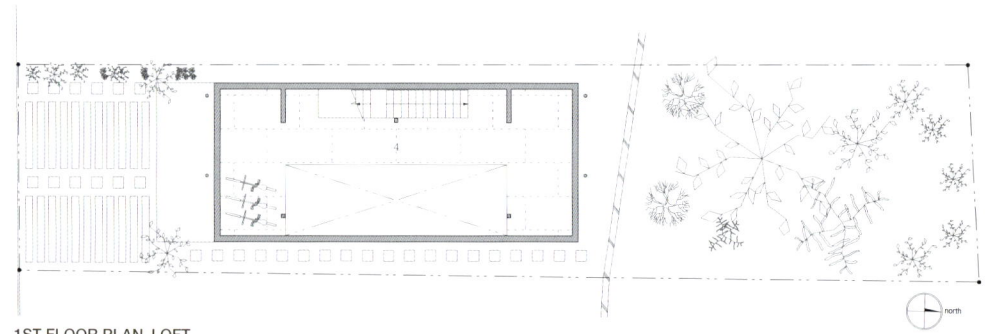

1ST FLOOR PLAN_LOFT

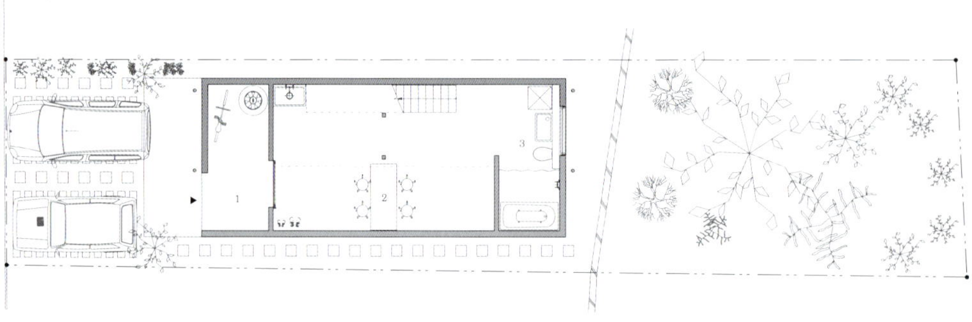

1ST FLOOR PLAN

1 ENTRANCE
2 ATELIER
3 BATH & TOILET
4 STORAGE

35m² - 41m²

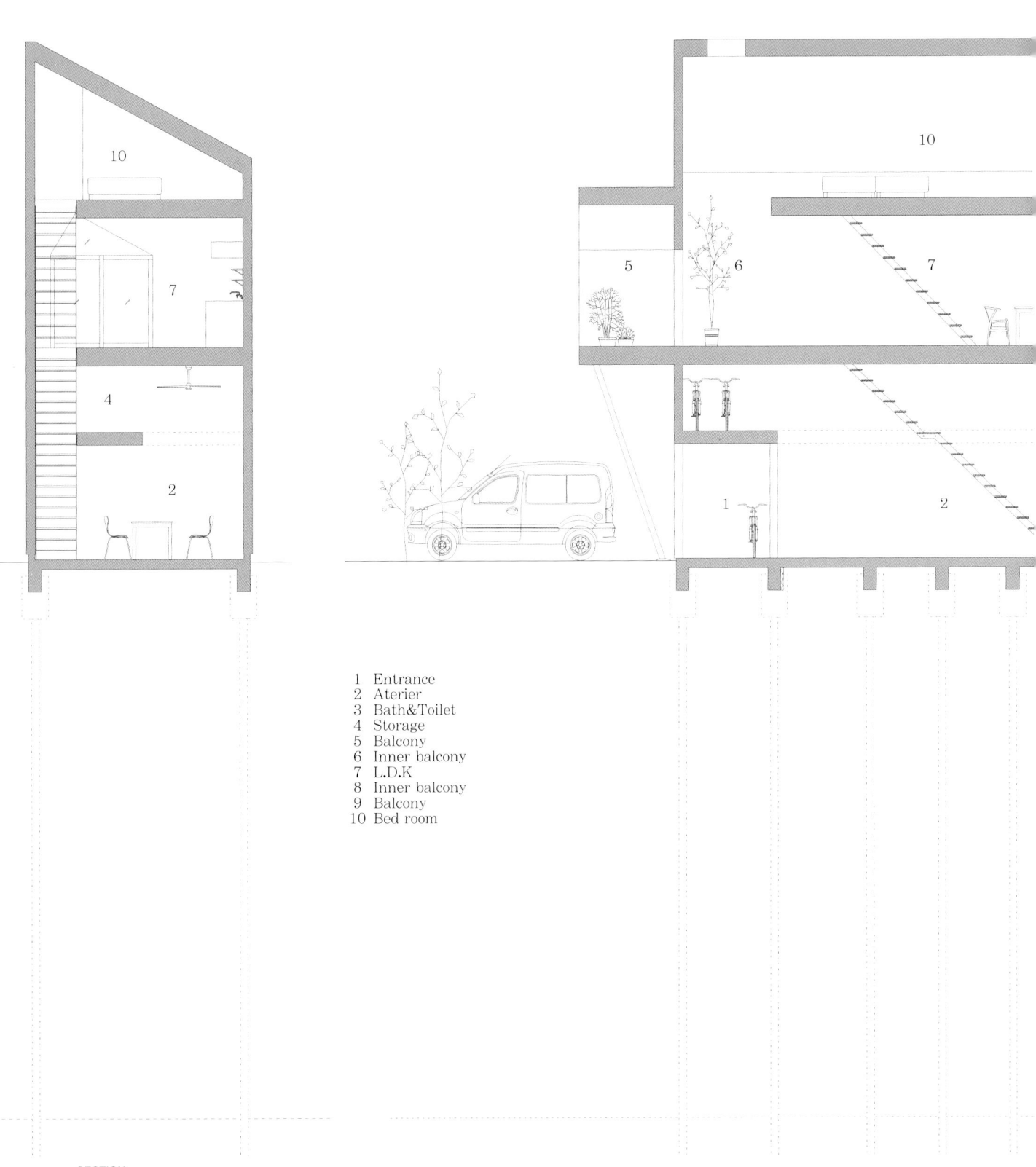

1 Entrance
2 Aterier
3 Bath&Toilet
4 Storage
5 Balcony
6 Inner balcony
7 L.D.K
8 Inner balcony
9 Balcony
10 Bed room

SECTION

35m² - 41m²

5 BALCONY
6 INNER BALCONY
7 L.D.K
8 INNER BALCONY
9 BALCONY
10 BED ROOM

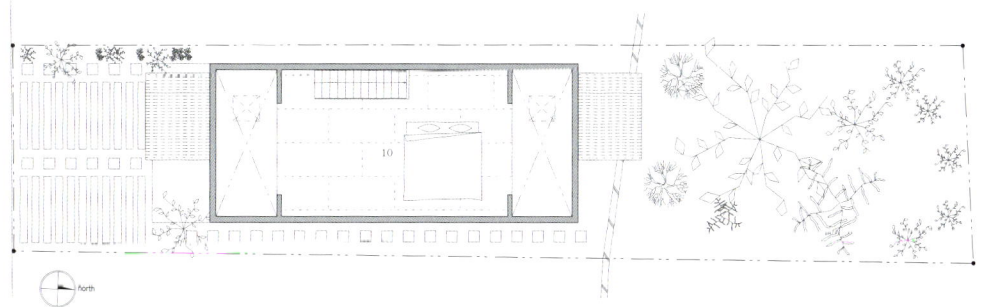

2ND FLOOR PLAN_LOFT

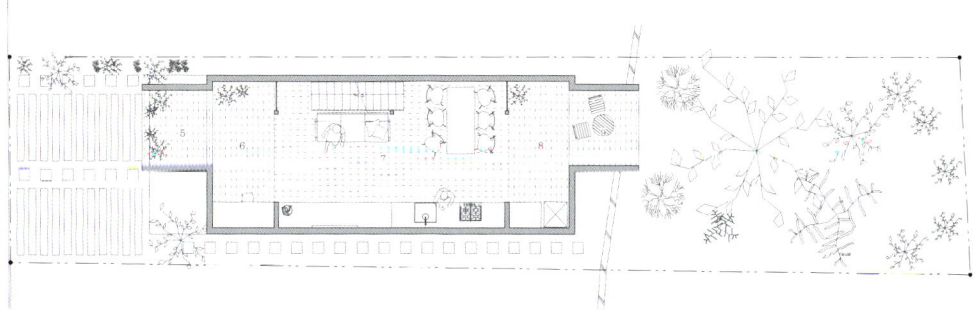

2ND FLOOR PLAN

no.555

interview

두 개 층 건물이지만 각 층에 중간층 역할을 하는 상단 공간을 두어 층을 수직적으로 분할하였다.

Q. 클라이언트의 요구 사항은 무엇이었으며, 그에 따른 건축적 대응은 어떤 것이었나?
A. 클라이언트는 취미활동을 위한 장비나 각종 물품을 가득 진열하거나 보관할 수 있는 공간을 원했다. 등산, 낚시 등 다양한 취미를 위한 장비들이 많았고 산악자전거만 해도 열 대나 있었다.

또한, 이런 장비를 손질하거나 수리하는 등 다양한 작업을 할 수 있는 공간을 필요로 하여 넉넉한 상단 적치 공간을 갖춘 작업실을 1층에 마련했다.
실내에는 심지어 화장실에도 문이 없을 정도로 개인적인 공간은 전혀 없다고 할 수 있다. 제목은 '집'이지만 실제로는 커다란 창고나 전시장이라고 할 수도 있을 것이다. 창을 통해 하나의 커다란 방으로 햇볕이 들어온다.

Q. 협소한 대지 위, 수납공간을 창출하기 위한 아이디어는 어떤 것이었으며, 수직적으로 프로그램 공간구성을 어떻게 하였는가?
A, 이 주택은 2층 건물이지만 각 층에 중간층 역할을 하는 상단 공간을 두었다. 각 층을 수직으로 분할한 셈이다. 따라서 각 층의 높이는 낮아졌지만 물품 보관이나 침실, 취미를 위한 공간을 보다 효율적으로 활용할 수 있도록 배치할 수 있었다.

Q. What were your client's requirements and your architectural approach for them?
A. Our clients requested spaces where they can store and decorate many hobby tools and collected items. They have many hobbies, mountain bike (10 bikes), mountain climbing, fishing and so on..
They also wanted a work space where they maintain and customs the tools. We made a atelier(all on the first floor) with loft(big storage).
There is no door in the house (even the toilet), it means there is no private room. The house is a house, but it is also a big warehouse or a showroom. The window and sun come through the big one room.

Q. What were your ideas to utilize the space & storage and how to zone the program in a vertical way?
A. The house is two floors building, however, each floor has loft space (intermediate floor). We layered the floors in a vertical way. The low height spaces can become an effective space (big storage, bedroom or hobby space...)

Contribution

another APARTMENT
Tsuyoshi Kobayashi
an-ap.com

1-p30, ROOFTOP HOUSE

another APARTMENT
Tsuyoshi Kobayashi
an-ap.com

1-p44, LONG WINDOW HOUSE

Atelier HAKO Architects
Tomomi Sano, Yukinobu Nanashima
www.hako-arch.com

1-p56, HOUSE AT HOMMACHI

CCPM Arquitectos
Constanza Chiozza,
Pedro Magnasco
www.ccpm.com.ar

1-p66, PH LAVALLEJA

Saunders Architecture
saunders.no

1-p78, SLICE

Kodasema
Ülar Mark
www.kodasema.com

1-p84, MINI HOUSE KODA

Taichi Mitsuya & Associates
Taichi Mitsuya
www.mtytic.com

1-p100, HOUSE IN KAWASAKI

a round Architects
(에이라운드 건축)
Park Changhyun
www.aroundarchitects.com

1-p112, YEONNAM-DONG MIX-USE HOUSING

Daisuke IBANO, Ryosuke FUJII, Satoshi NUMANOI
www.id-fr.com

1-p124, HOUSE IN THE CITY

Komada Architects' Office
Komada Takeshi, Komada Yuka
www.komada-archi.info

1-p140, TRANS

GONGGAM&KINFOLKS
(공감 건축사사무소)
Lee Yongeui, Song Kideok
www.kinfolks.kr

1 p152, H33617

APOLLO Architects & Associates
Satoshi Kurosaki
kurosakisatoshi.com

1-p160, NEST

ThePlus Architects
(건축사사무소 더함)
Cho Hanjun
the-plus.net

1-p170, [Crevice] 1740

Yuki Miyamoto Architect
Yuki Miyamoto
space.geocities.jp/ykm_arch

1-p184, SMALL HOUSE WITH FLOATING TREEHOUSE

Hiroshi Kuno + Associates
Hiroshi Kuno
qno.jp

1-P194, SIXTEEN ROOMS

People's Architecture Office(PAO)
He Zhe, James Shen, Zang Feng
www.peoples-architecture.com/pao

1-P206, Mrs. FAN'S PLUGIN HOUSE

a round Architects
(에이라운드 건축)
Park Changhyun
www.aroundarchitects.com

1-p218, SEONGSAN-DONG MIX-USE HOUSING

OOF! Architecture
Fooi-Ling Khoo
www.oof.net.au

1-P232, ACUTE HOUSE

PALMA
www.palmaestudio.com

1-P246, NARVARTE TERRACE

another APARTMENT
Tsuyoshi Kobayashi
an-ap.com

1-P258, CIRCULATE HOUSE

GONGGAM&KINFOLKS
(공감 건축사사무소)
Lee Yongeui, Song Kideok
www.kinfolks.kr

1-P272, S1927

APOLLO Architects & Associates
Satoshi Kurosaki
kurosakisatoshi.com

1-P280, HAT

GONGGAM&KINFOLKS
(공감 건축사사무소)
Lee Yongeui, Choi Yeonjung
www.kinfolks.kr

1-P288, H4912

Krupinski/Krupinska Arkitekter
KONRAD KRUPINSKI,
KATARINA KRUPINSKA
www.kkark.com

1-P298, SUMMERHOUSE T

Takeru shoji Architects
Takeru shoji, Yuki Hirano
www.takerushoji.jp

1-P308, Y HOUSE

RAUM Architects Group
(라움건축)
Oh Sinwook
rauma.co.kr

1-P318, VICOLO

no.555
Takuya Tsuchida, Kano Hirano
number555.com

1-P332, SUBAKO

Contribution

Takuro Yamamoto Architects
Takuro Yamamoto
takuroyama.jp

2-p30, LITTLE HOUSE WITH A BIG TERRACE

CHOP+ARCHI
Hiroo Okubo
www.chopweb.com

2-P44, KAMIUMA HOUSE

APOLLO Architects & Associates
Satoshi Kurosaki
kurosakisatoshi.com

2-P56, SLIDE

a21studio
www.a21studio.com.vn

2-P66, SAIGON HOUSE

AOC Architects
(에이오씨건축사사무소)
Hong Yangpyo
www.aoca.co.kr

2-P76, QUARTER HOUSE

ALTS DESIGN OFFICE
Sumiou Mizumoto
alts-design.com

2-P90, YAMASHINA HOUSE

B.L.U.E. Architecture Studio
Shuhei Aoyama, Yoko Fujii, Lingzi Liu
www.b-l-u-e.net

2-p98, DENGSHIKOU HUTONG RESIDENCE

Office of Architecture
Aniket Shahane
www.oa-ny.com

2-P108, LITTLE HOUSE. BIG CITY

S PLUS ONE architects
Yumiko Sakano
www.splusone.jp

2-P120, CEDAR HOUSE, PINE HOUSE

ŠÉPKA ARCHITECTS
JAN ŠÉPKA
www.sepka-architekti.cz

2-P128, HOUSE IN THE ORCHARD

AZO. Sequeira Arquitectos Associados
www.azoarq.com

2-P140, THE DOVECOTE

JYA-RCHITECTS
Won Youmin, Jo Janghee
jyarchitects.com

2-P156, YEONHUI-DONG 114

Yabe Tatsuya Architects
Tatsuya Yabe
www.somosomono.com

2-P168, YANA HOUSE

designband YOAP
(디자인밴드 요앞 건축사사무소)
Ryoo Inkeun, Kim Doran, Shin Hyunbo
yoap.kr

2-p182, PLAYFUL ATTIC HOUSE

Tsuruta Architects
www.tsurutaarchitects.com

2-P198, HOUSE OF TRACE

Frontofficetokyo
Will Galloway, Koen Klinkers,
Misuzu Yoshikawa, Joris Berkhot
frontofficetokyo.com

2-P268, OYAMADAI HOUSE

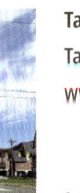
Takeru shoji Architects
Takeru shoji, Yuki Hirano
www.takerushoji.jp

2-P210, TOMI HOUSE

Rieuldorang Atelier
(리을도랑 아뜰리에)
Kim Seongyoul
www.rieuldorang.com

2-P288, NAMHAEJIB

Yabe Tatsuya Architects
Tatsuya yabe
www.somosomono.com

2-P220, KUTTE HOUSE

Hearth Architects
Yoshitaka Kuga
hearth-a.com

2-P300, SHOEI HOUSE

Apparte Studio
Otto Henkell
appartestudio.com.au

2-P236, CURTAIN COTTAGE

OpAD (오파드 건축연구소)
Oh Moonseok
blog.naver.com/opad_oms

2-P310, SOYUJAE

ERA architects
Esther Rovira
www.era.archi

2-P248, POLY HOUSE

Hearth Architects
Yoshitaka Kuga
hearth-a.com

2-P318, SANDAOSA HOUSE

OpAD (오파드 건축연구소)
Oh Moonseok
blog.naver.com/opad_oms

2-p260, DORIM-DONG MULTI-
DWELLING HOUSE 1+2

**Hiroto Suzuki Architects &
Associates**
www.sau.co.jp/~suzuki

2-P328, HOUSE IN WAKABAYASHI

협소주택 1
작은 면적, 넓은 공간

SMALL HOUSE 1
Small area, Big space

발행인 / 조배연
기획·편집 / 이선아, 국설희 (에뜰리에 www.etelier.kr)
디자인 / 신민기 (아뜰리에 파머), 에뜰리에
마케팅 / 정순안

인쇄 / 삼성문화인쇄
정가 / 88,000원
ISBN / 979-11-958268-8-9
출판등록번호 / 제2014-000167호

발행처 / 아키랩(월간 건축문화)
주소 / 서울시 서초구 양재천로13길 18(양재동)
전화 / 82-2-579-7747
이메일 / alarchilab@paran.com

*저작권법에 의하여 보호를 받는 저작물이므로 어떤 형태로든 무단 전재와 무단 복제를 금합니다.

Publisher / Cho Bae-yeon
Planning · Editing / Lee Sun-A, Kuk Seol-hee, (E'telier www.etelier.kr)
Design / Shin Min-ki (atelier farmer), E'telier
Foreign Business Dept. / Kevin Jung

Print / Samsung Moonhwa Printing Co., Ltd.
Price / USD 88
ISBN / 979-11-958268-8-9
Registration No. / 2014-000167

Publishing Office / ARCHI-LAB (Monthly review of Architecture & Culture)
Adress / 18, Yangjaecheon-ro 13-gil, Seocho-gu, Seoul, Republic of Korea
Tel / 82-2-579-7747
E-mail / alarchilab@paran.com

*All rights are reserved. Produced in South Korea. No part of this book may be reproduced in any form without written permission of the publisher.